Build AI-Enhanced Audio Plugins v

Build AI-Enhanced Audio Plugins with C++ explains how to embed artificial intelligence technology inside tools that can be used by audio and music professionals, through worked examples using Python, C++ and audio APIs which demonstrate how to combine technologies to produce professional, AI-enhanced creative tools.

Alongside a freely accessible source code repository created by the author that accompanies the book for readers to reference, each chapter is supported by complete example applications and projects, including an autonomous music improviser, a neural network-based synthesizer meta-programmer and a neural audio effects processor. Detailed instructions on how to build each example are also provided, including source code extracts, diagrams and background theory.

This is an essential guide for software developers and programmers of all levels looking to integrate AI into their systems, as well as educators and students of audio programming, machine learning and software development.

Matthew John Yee-King is a professor in the department of computing at Goldsmiths, University of London. He is an experienced educator as well as the programme director for the University of London's online BSc Computer Science degree.

"This book is long overdue. With the explosion of activity in the field of AI-assisted music creation, the need for mastering all the chain of software from ideas to actual plugins is stronger than ever. Matthew has a direct, hands-on approach that not only will be of great help to people wanting to contribute to the field, but will also encourage others to experiment and share their code. Matthew's experience in teaching shows and definitely contributes to making the book easy to read and to-the-point."

François Pachet, *Research Director*

Build AI-Enhanced Audio Plugins with C++

Matthew John Yee-King

Routledge
Taylor & Francis Group

LONDON AND NEW YORK

Designed cover image: Matthew John Yee-King

First published 2024
by Routledge
4 Park Square, Milton Park, Abingdon, Oxon OX14 4RN

and by Routledge
605 Third Avenue, New York, NY 10158

Routledge is an imprint of the Taylor & Francis Group, an informa business

British Library Cataloguing-in-Publication Data
A catalogue record for this book is available from the British Library

ISBN: 978-1-032-43046-1 (hbk)
ISBN: 978-1-032-43042-3 (pbk)
ISBN: 978-1-003-36549-5 (ebk)

DOI: 10.4324/9781003365495

Typeset in Computer Modern by Matthew John Yee-King

Access the Support Material: www.yeeking.net/book

Publisher's Note
This book has been prepared from camera-ready copy provided by the author.

For Sakie, Otoné, and my family. And of course, Asuka the beagle.

Contents

Foreword   x

List of figures   xi

I Getting started   1

1 Introduction to the book   2

2 Setting up your development environment   11

3 Installing JUCE   23

4 Installing and using CMake   32

5 Set up libtorch   44

6 Python setup instructions   53

7 Common development environment setup problems   60

8 Basic plugin development   62

9 FM synthesizer plugin   72

II ML-powered plugin control: the meta-controller   80

10 Using regression for synthesizer control   81

11 Experiment with regression and libtorch   87

12 The meta-controller   98

13 Linear interpolating superknob   103

14 Untrained torchknob   107

15 Training the torchknob   119

16 Plugin meta-controller   129

17 Placing plugins in an AudioProcessGraph structure   135

18 Show a plugin's user interface   143

19 From plugin host to meta-controller   151

III   The autonomous music improviser   157

20 Background: all about sequencers   158

21 Programming with Markov models   169

22 Starting the Improviser plugin   174

23 Modelling note onset times   187

24 Modelling note duration   194

25 Polyphonic Markov model   200

IV   Neural audio effects   209

26 Welcome to neural effects   210

27 Finite Impulse Responses, signals and systems   214

28 Convolution   220

29 Infinite Impulse Response filters   231

30 Waveshapers   241

31 Introduction to neural guitar amplifier emulation   254

32 Neural FX: LSTM network   261

33 JUCE LSTM plugin   274

34 Training the amp emulator: dataset   287

35 Data shapes, LSTM models and loss functions   296

36 The LSTM training loop   309

37 Operationalising the model in a plugin   315

38 Faster LSTM using RTNeural   320

39 Guide to the projects in the repository   328

Bibliography   335

Index   340

Foreword

I am delighted to present my book on building AI-enhanced audio software. My name is Matthew Yee-King, and I work as an educator, musician, and computer music researcher at Goldsmiths, University of London. I have written this book because I started making AI-powered audio plugins myself and found that musicians were much happier using plugins than other forms of software as they integrate with their existing tools. But there were no detailed instructions on how to integrate machine learning with audio in C++ for plugin development... until now! I hope you enjoy the book and find the techniques helpful for your work with AI-enhanced audio software. Here's my bio:

As an educator, I have taught undergraduate courses in digital signal processing, creative audio programming, software engineering, and artificial intelligence using languages such as Java, JavaScript, C++, and Python on-campus and online. I am the academic director for the first undergrad programme on the Coursera platform, the University of London's BSc Computer Science programme. In 2023, the course has seen over 8,000 students from more than 120 countries.

As a musician, I have released electronic music on Aphex Twin's Rephlex Records, Warp Records, and others. I have collaborated with many artists, including Tom Jenkinson (Squarepusher), Tom Skinner (Smile Band), Matthew Herbert, Finn Peters, Alex McLean, and Max de Wardener. My main live instrument is the drum kit, acoustic or electronic, but I have also worked as a sound designer, SuperCollider programmer, and creator of musical AI systems.

As a computer music researcher, I have developed music software and published papers on autonomous musical agents, automatic sound synthesiser programming, and systems supporting music education. I have worked on research projects as a research engineer, a post-doc, a co-I, and a PI. I have collaborated with individuals and research groups around the world, including Mark d'Inverno (my fantastic mentor), Andrea Fiorucci, Francois Pachet, Jon McCormack, Mick Grierson, Nick Collins, Rebecca Fiebrink, the Sony Computer Science Laboratory Paris, the Artificial Intelligence Research Institute, CSIC Barcelona, the Department of Human Centred Computing, Monash University Australia, and Politecnico di Milano, Italy.

I would like to thank all the amazingly talented musicians (especially Gaz, Domenico, Finn, Alex, and Max) and researchers I have worked with for inspiring me and, indeed, the work on which some of the projects in the book are based.

List of figures

1.1 Mountainous landscape depicting musical competencies, with the water of AI rising to consume them. 2

2.1 Many component parts are needed to build AI-enhanced audio software. 11
2.2 The components involved in AI-enhanced audio application development. 12
2.3 Setting up your development environment involves complex machinery and lots of steps. 18
2.4 Creating, building and running a C++ console program in Visual Studio. 19
2.5 C++-related packages that you should install. 19
2.6 Creating, building and running a C++ console program in Xcode. 20
2.7 My development setup showing an M1 Mac running macOS hidden on the left, a ThinkPad running Ubuntu 22.04 in front of the monitor and a Gigabyte Aero running Windows 11 on the right. . 22

3.1 The available application types for Projucer. 24
3.2 Projucer project view. The exporter panel is exposed on the left, module configuration panel is on the right. Note that my modules are all set to Global. 25
3.3 Building the Standalone solution in Visual Studio Community 2022. 27
3.4 Enabling console output for a JUCE project in Visual Studio/ Windows. At the top: redirect text output to the immediate window and open the immediate window. At the bottom: a program running with DBG output showing in the immediate window. 28
3.5 Running a JUCE plugin project in Xcode – make sure you select Standalone. 29
3.6 The JUCE AudioPluginHost application, which comes with the JUCE distribution. One of its built–in plugins, a sine synth, is wired to the MIDI input and the audio output. 30

3.7 The list of available plugins in the JUCE AudioPluginHost app. I have clicked the options menu which is showing its 'scan for new or updated ...' function. 31

4.1 CMake running in the Windows Powershell. 33
4.2 A CMake project viewed in VSCode. 36
4.3 A CMake project viewed in Visual Studio Community 2022. . . . 38
4.4 A CMake project viewed in Xcode. 39

6.1 A Jupyter notebook in action. There are cells containing Python code which you can execute. If you trigger a plot command, the plot will be embedded in the worksheet. 57

8.1 dphase depends on the sample rate (the space between the samples) and the frequency (how fast you need to get through the sine wave). 64
8.2 A synthesizer plugin loads data into the incoming blocks. 65
8.3 Printing descriptions of MIDI messages coming into a plugin in Standalone mode from a USB controller keyboard (left) and MIDI coming from an on-screen piano keyboard in AudioPluginHost (right). 70

9.1 Simple FM plugin with sliders for frequency, modulation index and modulation depth. 75
9.2 Showing plugin parameters for the Surge XT synthesiser using AudioPluginHost. 76
9.3 Showing plugin parameters for the FM plugin using AudioPluginHost. 77
9.4 Showing the custom UI for the FM plugin (right), the auto-generated parameter UI (middle) and AudioPluginHost (left). . . . 78

10.1 The meta-controller uses regression to control other plugins. 81
10.2 Linear regression finds the straight line that best fits some data. Important features of the line are the point at which it intercepts the y-axis and the slope gradient. 82
10.3 Linear regression with two lines, allowing the estimation of two parameters given a single 'meta-controller' input control. The x-axis represents the control input and the y-axis shows the settings for the two parameters the control is mapped to. 84
10.4 A neural network applies a function to its input. 85
10.5 A neural network scales by a weight and adds a bias. 86

11.1 Simple single layer network with one input and one output. 91

11.2 More complex single layer with more inputs and outputs. Now we apply a weight to each input as it goes to each output. We then sum the weighted inputs and apply a bias to each output. 93

11.3 The optimiser adjusts the network weights. 95

12.1 The meta-controller uses a neural network for new methods of synthesizer sound exploration. 98

12.2 The Wekinator workflow: data collection, training, inference then back to data collection. 101

13.1 User interface for the simple two-parameter FM synthesizer. The toggle switch switches between drone and envelope mode, the two sliders control modulation depth and index and the piano keyboard allows you to play notes on the synthesizer. 103

13.2 Superknob UI on the left. On the right is a closer view of a range slider. Small triangles above and below the line allow the user to constrain the range of the main slider control. 104

14.1 The torchknob system architecture. 108

14.2 Basic architecture where a linear layer passes into a softmax layer. The numbers in the brackets indicate input and output shape. The linear layer input (2,1) goes from 1 value to 2 nodes, then output (2,2) goes from 2 nodes to 2 outputs. 115

15.1 Interactive machine learning provides more intuitive training for neural networks. 119

15.2 User interface mockup for trainable superknob system (left). We have an additional knob to specify training input without triggering the movement of the sliders. Actual user interface prototype (right). 120

15.3 Example of an experiment you can carry out. First, set the training slider to its lowest value, the same for the modulation controls. Add a training point. Then move to the middle positions, and add a training point. Finally, move to the highest positions, and add a training point. 125

15.4 The learner.js/ Wekinator regression architecture (top). The simpler architecture we used previously (bottom). 125

16.1 Hosting plugins allows you to control more advanced synthesizers. 129

17.1 Wiring plugins together with a processor graph. 135

17.2 The graph you will create. 139

17.3 User interface for the basic host with the load plugin button added. 141

18.1 Class hierarchy for AudioProcessor and its descendants. 143
18.2 User interface for the host with the Surge XT plugin user interface showing in a separate window. 144
18.3 User interface for the host with a show UI button. 145

19.1 User interface for the Dexed DX-7 emulator. It has 155 parameters. 151
19.2 Time for a more complex neural network. 153

20.1 How far are modern sequencers from steam-powered pianos? 158
20.2 My own experience interacting with AI improvisers. Left panel: playing with Alex McLean in Canute, with an AI improviser adding even more percussion. Right panel: livecoding an AI improviser in a performance with musician Finn Peters. 159
20.3 Visualisation of a two-state model on the left and the state transition probability table on the right. 164
20.4 Visualisation of a variable order Markov model containing first and second order states. 166

21.1 Example of the Markov model generated by some simple code. . . 170

22.1 Overview of the autonomous improviser plugin. Yes, a keytar. . . . 175
22.2 The user interface for the basic JUCE MIDI processing plugin. . . 176
22.3 The MIDI Markov plugin running in AudioPluginHost. Note how it receives MIDI and then passes it on to the Dexed synthesiser. . . 181
22.4 Using AudioPluginHost's MIDI Logger plugin to observe the MIDI coming out of the Markov plugin. 182

23.1 Note duration is the length the note plays for. Inter-onset interval is the time that elapses between the start of consecutive notes. . . 187
23.2 Measuring inter-onset-intervals. The IOI is the number of samples between the start and end sample. elapsedSamples is the absolute number of elapsed samples since the program started and is updated every time processBlock is called; message.getTimestamp() is the offset of the message in samples within the current block. 188

24.1 Measuring note duration has to cope with notes that fall across multiple calls to processBlock. 194
24.2 Testing the getTimestamp function on note–on messages – the timestamp is always between zero and the block size of 2048. . . . 196

25.1 If notes start close enough in time, they are chords. If the start times fall outside a threshold, they are single notes. This allows for human playing where notes in chords do not happen all at the same time. 202

26.1 Tape manipulation was an early form of audio effect. 210

27.1 The impulse signal and the impulse responses of a one-pole, two-pole and three-pole system. 216

28.1 Original drum loop spectrum on the left, filtered version on the right. High frequencies have been attenuated in the filtered spectrum. 224

29.1 Infinite impulse responses are powerful! 231
29.2 Pole for pole, IIR filters generate much richer impulse responses. The left panel shows a two-pole, FIR. The right pane shows a two-pole IIR. 232
29.3 Comparison of two pole FIR filter (left) and IIR filter (right). The IIR filter has a more drastic response. 233
29.4 Two types of IIR filter and their frequency responses. IIR filter design is a compromise. 237

30.1 Digital signal processing makes waveshaping much easier than it used to be. 241
30.2 The effect of different waveshaper transfer functions on a sinusoidal signal. Top row: transfer functions, middle row: sine wave signal after waveshaping, bottom row: spectrum of waveshaped sine wave. 242
30.3 Automatically generated generic UI for the waveshaper plugin. . . 248
30.4 A sine wave passing through a series of blocks that emulate in a simplified way the processing done by a guitar amplifier. 249
30.5 Capture an impulse response for the convolutional cabinet simulator. 253

31.1 Capturing training data from a guitar amplifier. 256
31.2 Four stages to train a neural network. 1: send the test input through the device (e.g. amp) you want to model, 2: send the test input through the neural network, 3: compute the error between the output of the network and amp, 4: update network parameters to reduce error using back-propagation. Back to stage 2. 257

32.1 What does our simple, random LSTM do to a sine wave? It changes the shape of the wave and introduces extra frequencies. 266

32.2 The steps taken to process a WAV file with a neural network through various shapes and data formats. 270

32.3 Time taken to process 44,100 samples. Anything below the 1000ms line can potentially run in real-time. Linux seems very fast with low hidden units, but Windows and macOS catch up at 128 units. . . 272

33.1 Block-based processing leads to unwanted artefacts in the audio. The left panel shows the output of the network if the complete signal is processed in one block. The right panel shows what happens if the signal is passed through the network in several blocks. The solution is to retain the state of the LSTM between blocks. 277

33.2 Breakdown of the data type used to store LSTM state. 280

33.3 The LSTM plugin running in the AudioPlugHost test environment, with an oscilloscope showing a sine wave test tone before and after LSTM processing. 284

34.1 How fast can LSTMs process audio? 287

34.2 Four stages to train a neural network. 1: send the test input through the device (e.g. amp) you want to model, 2: send the test input through the neural network, 3: compute the error between the output of the network and amp, 4: update network parameters to reduce error using back-propagation. Back to stage 2. 288

34.3 Tensorboard is a web-based machine learning dashboard. Here, you can see a list of training runs (1) and graphs showing training progress in terms of training (2) and validation (3) errors on two separate runs. 291

34.4 Spectrogram of the Atkins training signal 'v2_0_0.wav'. You can see the signal is quite varied and dynamic. 292

34.5 Capturing training data from a guitar amplifier is similar to 're-amping'. 292

34.6 Clean signal (top) and re-amped signal (bottom) in Reaper. . . . 294

35.1 Sequence length and batch size. 298

35.2 An LSTM network with a four hidden unit LSTM layer and a densely connected unit which 'mixes down' the signal to a single channel. 303

35.3 What does loss mean? The top two plots show extracts from the target output. The middle two plots show the output of an untrained (left) and trained network (right). The right-hand side is much closer to the target. The bottom plots show a simple error between each point in the two plots above. The sum of these values could be a simple loss function. 305

36.1 The training loop. Data is processed in batches with updates to the network parameters between batches. Between epochs, checks are done on whether to save the model and exit. 310

36.2 Comparison of training runs with different sized LSTM networks. At the top you can see the input signal and the target output signal recorded from Blackstar HT-1 valve guitar amplifier. The descending graphs on the left show the validation loss over time for three LSTM network sizes. The waveforms show outputs from the networks before and after training. 314

38.1 A sinusoidal test signal passing through an RTNeural LSTM distortion effect in AudioPluginHost. The sinusoidal wave is the original signal, the clipped out wave is the LSTM-processed signal. 325

Part I
Getting started

1

Introduction to the book

Welcome to **'Build AI-Enhanced Audio Plugins with C++'**! You are about to embark on a journey into a world of advanced technology, which will be an essential part of the next generation of audio software. In this chapter, I will introduce the general area of AI-music technology and then set out the book's main aims. I will identify different types of people: audio developers, student programmers, machine learning engineers, educators and so on and explain how each group can get the best out of the book. I will explain that you can use any of the large amounts of code I have written for the book however you like. I will also explain the dual licensing model used by the JUCE library. I will finish with a straightforward, working definition of artificial intelligence.

1.1 Exciting times for artificial intelligence

FIGURE 1.1
Mountainous landscape depicting musical competencies, with the water of AI rising to consume them.

As I write this book, we are in exciting times for the progress of artificial intelligence (AI). AI theorist Hans Moravec presents a compelling view of the progress of AI wherein he places human competencies such as picking up a cup, composing a symphony and so forth within a mountainous landscape he calls a "landscape of human competencies"[29]. Complex or highly valued competencies are situated on higher peaks, although people may disagree on what should go where.

In Moravec's landscape of human competencies, artificial intelligence is a sea washing around and rising up the peaks. If the water rises above the position of a human competency, AI has that competency. Right now, there is no doubt that the sea is rising, especially given recent advances in deep neural networks. Game playing is a popular area for AI researchers, and it can provide some perspective on the position of the water level. Chess is underwater[20],

Go is underwater[38], and so is Heads-up no-limit Texas hold'em Poker[5]. But picking up cups is still hard at the time of writing. This leads us to another insight from Moravec known as Moravec's Paradox. Things that people find hard are often easy for AI, and things that people find easy can be very hard for AI.

What about music and sound engineering competencies? Well-known baroque composer Bach is swimming in the sea of AI after Hadjeres's DeepBach created Bach chorales that could fool experts[14]. AI can learn how to program the notoriously difficult to program Yamaha DX-7 synthesizer to make a given sound[49], and AI can automatically mix multi-track audio recordings[28]. AI has even gained competencies which humans do not have. Timbre transfer allows us to use the pitch and dynamics of one instrumental performance to play using the tone of another instrument[12], and sound separation allows us to un-mix a track back to its separate stems[41]. Recent advancements in voice processing technology have allowed 'deep fakes' wherein the voices of dead musicians can be re-animated as if in a kind of Lovecraftian fantasy and made to sing cover versions[1].

Many of those examples are from the research literature about AI and music but commercial music software companies are increasingly using AI in their products. A notable example is Waves and their neural network-powered noise reduction technology Clarity Vx. Another example is Steinberg's GANDrum system, which uses generative adversarial neural networks to synthesize unique drum sounds. There are also an emerging range of neural network powered guitar effects such as Neural DSP's Quad Cortex guitar amplifier emulator.

So right now is an exciting time to be working as a music technology developer because a revolution is underway which is likely to be at least as transformative as the desktop Digital Audio Workstation revolution from the late 90s, which put advanced recording and audio processing capabilities in the hands of anyone who owned a desktop computer. In fact, AI is even more exciting than DAW technology because it is transforming the way we produce and create music in ways that were previously unimaginable. The possibilities are profound, and I am excited to see what the future holds.

1.2   The aims for this book

The first aim of this book is to show you how to build AI-enhanced music software. But I will take a different approach to many other AI and signal processing books. I have carefully developed this approach to solve specific problems and to address particular challenges you will face as a developer of AI-enhanced music software.

---

[1]https://github.com/svc-develop-team/so-vits-svc

One problem with many AI-music systems I mentioned is that they only exist as descriptions in research papers. The research papers aim to describe systems and their performance to other AI-music experts using a very limited number of pages. They are not tutorials and do not necessarily explain the nuts and bolts needed to build a complete working system.

Sometimes the researchers who write these research papers provide source code repositories but my experience has been that it can be challenging to operationalise the software from these source code repositories. Making the code run likely involves having a particular combination of particular versions of other components installed and often a particular operating system. The challenge is even more significant if there is no source code repository. I have watched excellent PhD students labour for weeks to re-implement systems described in research papers, only to discover there are vital details that should have been included in the paper or other technical issues.

These problems mean that AI-enhanced music systems are not easily accessible to musicians wishing to use them and probably not to programmers wanting to integrate them into innovative music software. This is where this book comes in. This book will show you several examples of how to build complete working AI-music systems. All source code written by me is provided in a repository and is covered with a permissive open-source license, allowing you to re-use it how you see fit. The book also uses a consistent technical setup, allowing you to easily access and "wrench on" the examples. So my first aim is really to make AI-enhanced music technology available and transparent for you.

My second aim is to show you how to construct the technology in a way that makes it accessible to musicians and audio professionals. Knowing how to build and run AI-enhanced music systems on *your* machine is one thing. A quirky Python script hacked together to work on your setup is acceptable for research and experimentation but you will have trouble getting musicians to use it. What is the ideal method for sharing software in a form that musicians and other audio professionals can use it? The next aim of this book is to answer that question and to apply the answer to the design of the examples in the book. For many years I have worked as a researcher/engineer on research projects where one of the aims has been to get new technology into the hands of users so we can evaluate and improve that technology. There are many approaches to achieving this aim: running workshops with pre-configured equipment, making the technology run in the web browser and so on. All of these approaches have their merits and are appropriate in different circumstances. But none of them is quite suitable for our purposes here. Here we are aiming to write software that can be used by musicians, producers and sound engineers with minimal effort on their part.

How can you achieve this aim of having as many audio professionals as possible to be able to use your software, and how can this book help you? Firstly, making software so it integrates with existing creative workflows is crucial. The simplest way to do that in music technology is via standard plugin frameworks such as VST and Audio Units or standalone, native applications that work with standard protocols such as MIDI and OSC. In this book, we will do precisely this by building software that integrates effectively with existing workflows and technology.

1.3 What is in the book?

I have organised the book into four parts:

1: Getting started. You will set up your system for the development work in the book and build some example plugins and other test programs.

2: ML-powered plugin control: the meta-controller. You will build the first large example in the book, a plugin that hosts and controls another plugin using a neural network.

3: The autonomous music improviser. The second large example is a plugin that can learn in real-time from incoming MIDI data and improvise its own interpretation of what it has learned.

4: Neural audio effects. The third large example is a plugin that models the non-linear signal processing of guitar amplifiers and effects pedals using neural networks.

Each part of the book contains detailed instructions on how to build each example, including source code extracts, diagrams and background theory. I have also included some brief historical and other context for the examples. The examples in parts 2, 3, and 4 are independent, so you can jump to any of those parts once you have completed part 1.

I have created a freely accessible source code repository, currently on Github, which provides each of the examples above in various states of development. As you work through each stage of developing each example, you can pull up a working version for that stage from the repository, in case you get stuck. The repository contains releases of each final product with compiled binaries and installers.

1.4 How to use this book

There are different ways to use the book. The most straightforward approach is to download the releases of the examples and experiment with them in your DAW.

But you will miss out on a lot of learning if you only do that! To fully exploit the content in the book, you will need to start by working through part 1, which explains how to set up your system for the development work described in the book. You will find 'Progress check' sections at the end of each chapter, which clarify what you should have achieved before continuing to the next chapter.

After part 1, there are three detailed example projects, each providing very different functionality. These three projects are independent, so you can choose which order you study them in or only study some and not others.

As you work through the parts of the book, each program you are developing will increase in complexity until it is completely functional at the end of the book part. To work through the examples, you can type in all the code you see in the book and build the complete example by hand, or you can read the code and download the step-by-step versions of the projects from the repository. I find that people sometimes get really stuck working through these larger projects, where they cannot make it compile or work properly. So, I have provided staged versions of the programs in the code repository. If you reach the end of a chapter and cannot figure out why your program does not work, you can just pick up from a working stage in the code repository and continue. Of course, there is much to be gained by spending hours looking for that missing bracket, so do try and debug your problems before grabbing the working version from the repo.

At the end of some chapters and at the end of all parts, I suggest challenges and extensions. These are extra features you can add to the plugins, allowing you to reinforce and increase your understanding of the principles and techniques. If you are using the book for teaching, these challenges and extensions are things you could set students for coursework.

1.5   Who might use this book?

In the following sections I will mention a few different types of people who would be interested in the book and how they can get the best out of it.

1.5.1   Student programmer

If you are a student programmer at the undergraduate or postgraduate level, you will find new knowledge in all areas of the book. You might be assigned all or part of this book as a textbook for a course you are studying, or you might have discovered the book independently. The book will teach you some C++ programming along the way, as well as helping you to understand how to use an IDE and associated tools to develop software. You will probably find the setup section of the book very helpful as, in my experience, students spend a lot of time struggling to get their software development environment set up and working correctly. The book also provides information about general audio processing techniques and how they can be adapted using AI technology.

1.5.2   Audio programmer

You will find the book helpful if you are an audio programmer wishing to integrate AI technology into your software. The book will show several detailed examples of how you can employ different AI techniques in music software. The code repository contains permissively licensed code which you can use how you see fit in your projects. The book considers cross-platform development and is focused on developing software that integrates with existing music production workflows and technology. You will probably be familiar with the audio side of the development and theory work in this book, and I will try to make connections between this knowledge and the AI domain. For example, you will discover how neural networks are just big fancy signal processors.

1.5.3   Machine learning engineer

If you are a machine learning engineer or AI scientist wishing to apply your skills in the music software domain, the book will help you do that. You will be familiar with the concepts of machine learning models, training and inference, but you will likely need to become more familiar with digital signal processing techniques. You may be unfamiliar with common music technology such as plugins, MIDI data, etc. The book will explain exactly how to work with those technologies. As a machine learning engineer, you likely work primarily in Python or a specialised language. The book covers some Python but is mainly focused on C++ programming. You should find the information about setting up a C++ development toolchain useful.

1.5.4   Educator

If you are an educator planning to use this book as part of your teaching, that is an excellent idea! The most obvious way to use the book in your teaching is to split the content between lectures and lab classes. In the lectures, you can introduce the AI theory. You can go as far as you like with the AI theory, depending on the level and focus of your course. I cover enough in the book to enable the reader to carry out training and inference and to integrate the AI system into a working application. I also explain some characteristics of the particular machine learning techniques used. You can go much further than that, depending on your requirements. For your lab classes, your students can work through the practical implementation of the applications. The book contains detailed instructions on how to build each of the example applications. I have battle tested and iterated these instructions with my students.

1.5.5 Sound engineer/ sound designer/ musician

You might be a sound designer or musician who wants deeper control and knowledge of AI-enhanced music systems. In that case, you probably need to become more familiar with the world of software development or machine learning and AI, and the book will help you to achieve that. You will have strong domain knowledge covering the musical aspects of the systems developed in the book. You should start by checking out the example software available in the repository. If you are intrigued by those examples, you can use the book to learn more about how they work. That can lead you to customise the software to suit your needs or ideas better.

1.6 The source code repository

The book comes with a source code repository on GitHub. The source code repository contains all the code for the projects described in the book and various instructional materials. You should go ahead and install the git tool and clone the repository to your machine right now. Chapter 39 describes each of the projects you will find in the source code repository. I refer to this chapter when I am working through example projects to ensure you can access the correct code for that project directly. This is a book, and books cannot easily be updated, but the software libraries we are working with are regularly updated. This causes tension between the desire to provide the most correct and up-to-date instructions and code in the book and wanting to ensure the instructions remain correct. If you find any bugs in the example code in the repository or in the book, please report them as issues on the GitHub page. I cannot do much about updating the code in the book (pending a new edition), but I will ensure the repository is as bug-free as possible.

1.6.1 About source code licensing

In this section, I will explain the licensing model used for code in this book. When I read these kind of legal details of source code licensing, my eyes glaze over, and I start wishing for the paragraph to end. But please read this section, as it is important to understand how to use the code here. Here is the executive summary: you can use any of the code I have written for this book freely in your commercial projects but if you want to use the complete examples in a closed-source manner you should ensure you understand the JUCE library's dual licensing model.

Open-source has various definitions. For our purposes, open-source means that the source code for a piece of software is available. Open-source code generally comes with a license that dictates what the code's author wants you to do with that code. Permissive licenses, such as the MIT license, place few limitations on the use of that code. Users are free to use that code as they see fit. Users of MIT-licensed code can adapt the code and even include it in commercial projects without needing to release their adapted code. All the code written by me for this book is MIT licensed.

The GPL licenses take a stronger philosophical position concerning freedom and are designed to encourage further sharing of source code. You can adapt and use GPL'd code in your project, even if it is a commercial project, but you will be required to release your code. You are also obliged to make your source code open-source with a GPL license if your code links to GPL libraries.

Releasing source code with multiple licenses is possible if they are compatible. The code in this book that I have written is released under a dual MIT / GPL license. I will explain why below. If you use my code in your project, the MIT license applies. The GPL license applies if you use my complete examples, which also link to GPL'd code.

The reason I have dual-licensed the code is because of the JUCE library. The JUCE library carries a dual license. If you build against the JUCE library, you can either GPL your whole project or apply for a JUCE license, and then, you do not need to GPL your code.

1.6.2 Example code-use scenarios

Here are some examples to illustrate typical ways you might want to work with the source code from this book.

1. I want to experiment with the examples in this book. I am not planning to release anything commercially. Great – you can do that, no problem.

2. I want to use the code written by the author of this book in my own plugin projects, where I am using my own framework instead of JUCE. So I am not going to build against the JUCE library. Great – the MIT license applies to my code in that scenario, and you are free to release projects which use that MIT code commercially.

3. I have adapted an example in the book into a really cool plugin I want to release commercially. I do not want to release my complete code with a GPL. It uses the JUCE library. Do not worry; you can do that. You need to get yourself an appropriate license from the JUCE web page.

1.7   User-readiness

I will show you how to get the software to a point where it will run on your machine and, with some fiddling, on other people's machines. I do not cover the creation of installers or the process of signing / notarising software. Some great resources online tell you what to do with installers, signing etc., once your amazing AI-powered plugin is ready for the world.

1.8   A working definition of artificial intelligence

To complete the pre-amble here, I would like to provide a working definition of what I mean when I refer to artificial intelligence and how that differs from machine learning and other software development techniques. Here goes:

An artificially intelligent system is an automated system that can carry out a task generally considered to require intelligence were it to be carried out by a human. Machine learning refers to a set of techniques that can be used to build artificially intelligent systems amongst other things. Machine learning techniques involve learning in the sense that the program changes itself or its parameters in a manner that allows it to perform a task more and more effectively. Often machine learning involves learning patterns in data.

2

Setting up your development environment

In the next few chapters, I will explain how you can set up your development environment for audio software development. You need this setup to work on and run the example programs in the book. After working through these chapters, you should be able to build and run a simple C++ program linked to the JUCE audio library and the libtorch machine learning library using an integrated development environment (IDE). The chapters should also familiarise you with the CMake tool which will allow you to create cross-platform projects with which you can create native applications and plugins for Windows, macOS and Linux systems. You can use these setup chapters how you like – read them, scribble on them, etc. but I recommend working through the material with a computer available. Expect to install software on the computer, run commands in its command shell and execute programs.

2.1   Component parts

FIGURE 2.1
Many component parts are needed to build AI-enhanced audio software.

Before we set up the development environment, I would like to describe some key components you will encounter in this environment. You are probably familiar with several of these components, but I am describing them here to clarify what I plan to use them for. Several of these components are shown in relation to each other in figure 2.2.

2.1.1   Purpose of different components

Build tool

A build tool helps developers specify how their software should be built. For example, what are the software libraries they are using? Where are the source code files? What is the target platform? Build tools are handy when you want

11

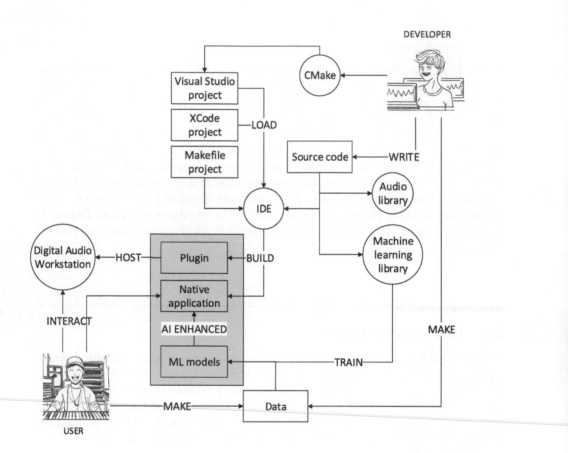

FIGURE 2.2
The components involved in AI-enhanced audio application development.

to be able to build your software for multiple platforms using different Integrated Development Environments (IDEs). We will use the CMake[1] build tool to help us generate projects for various IDEs. This will make building the software for different hardware and OS platforms possible.

CMake

CMake is the build tool we will use in this book. With CMake, you write a single configuration file then you can use it to generate projects for different IDEs. In the configuration file, you can specify different targets for the build, such as a test program and a main program. You can specify associations between your project and external libraries. You can specify actions to be taken, such as copying files. This makes it a valuable tool for audio application developers wishing to support various operating systems, as you can maintain a single CMake configuration and codebase and use it to build for multiple platforms.

Integrated Development Environment (IDE)

An IDE is a set of tools that enable a developer to write, build and debug software. The IDE will be the hub of your software development activity. Using the CMake build tool, you can generate projects for different IDEs from the same codebase. Many IDEs are available, but I will cover Xcode for Apple device development, Visual Studio for Windows and Visual Studio Code or a custom setup for GNU/Linux in this book. In fact, you can also use Visual Studio Code for Windows, Apple and Linux development, which allows for a consistent environment across platforms.

Codebase

The codebase is the set of source code files in a project. The build tool and some handy macros will allow us to have a single codebase for all platforms.

Native program

Some of the programs you encounter in the book will be compiled into native programs in machine code that run on particular CPU hardware. We will write these programs using the C++ language. Native programs are most appropriate when developers wish to integrate directly with plugin and Digital Audio Workstation (DAW) technology. For example, VST3 plugins are native programs. Native programs generally run faster than interpreted programs, and that is important for realtime audio applications.

---

[1] https://CMake.org/

Interpreted program

Some of the programs we write will be interpreted as opposed to compiled. Interpreted programs are converted to machine code on the fly instead of being converted into machine code before running. We will write interpreted programs in the Python language. Interpreted programs are more suited to the kind of experimentation one needs to do when developing machine learning models. It is common for AI researchers to provide Python code along with their research papers to allow other people to explore their work more easily. It is less common for researchers to provide C++ code, but there has been a trend towards this in AI-music research in the last few years.

Machine learning model

Machine learning models carry out the smart processing associated with artificial intelligence systems. A neural network is an example of a machine learning model. You can think of a machine learning model as a kind of data processing black box that can learn to process data in a way that is useful to us. The structure of the model (inside the black box) and its parameters dictate what kind of processing it does. Depending on the application, we will sometimes develop the models using Python and sometimes C++. Experimenting with model designs and training data in Python is generally more straightforward. C++ allows the models to be integrated into native applications and plugins so they can be accessed by regular users.

Trained model

A trained model is a machine learning model that has learnt something valuable. The model defines the structure of the machine learning component. Training teaches that component to process data in a particular way. The ability to self-configure through learning is the essence of machine learning. Trained models provided by a third party are often called pre-trained models. In case you have heard of openAI's infamous GPT model, the 'P' stands for pre-trained.

Inference

Inference is the process of using a trained model to generate an output. Inference does not change the internal configuration of the machine learning model; it just passes data through it. Once a model is trained, you will use it for inference.

Generative model

A generative model is a machine learning model that can generate something interesting. For example, instead of detecting cats in images, it might generate images of cats. The 'G' in GPT stands for generative, as GPT generates text.

Machine learning library

A machine learning library is a set of components that makes it easier to carry out machine learning tasks. Typical components include different algorithms such as clustering and neural networks, routines to train the models, and data importers. Examples are sci-kit learn, TensorFlow and PyTorch. We will use the PyTorch library along with its C++-compatible library libtorch. This will allow us to work in both Python and C++, taking advantage of the strengths of each language. libtorch is also compatible with the CMake build tool.

Audio library

An audio library is a set of components that makes it easier to construct audio applications. Typical components are audio file readers and writers, audio device management and sound synthesis routines. We will use the JUCE audio library as it allows us to construct cross-platform audio applications and plugins. If you are not keen on using JUCE, you should be able to convert the applications we make to work in other audio libraries. For example, it is possible to create VST plugins directly using Steinberg's library, or you could use IPlug2. There are a few reasons I have chosen to use JUCE: it provides a very consistent experience on different platforms, it includes a set of cross-platform user interface components; it is compatible with the CMake build tool, and it can export plugins in several formats such as VST3, AudioUnit and so on. JUCE also provides components to build applications and plugins that can host other plugins, a capability we will make use of in some of the examples in the book.

Application Programming Interface (API)

An API is a set of ready-made components which you can use to develop applications. APIs come in different forms but the JUCE API contains C++ classes representing user interface components, audio file readers and so forth. See also the comments about libraries below.

Dynamic and static linking

When you build native applications, part of the process involves connecting your application somehow to the libraries you have used. There are two options here: dynamic linking and static linking. Static linking means the library (for example, the audio library) is included inside your program's binary. Dynamic linking means the library exists as a separate file, and your application stores a reference to it. Depending on your platform, dynamically linked libraries are also called DLLs, shared objects and dylibs. When you run an application with dynamic links, those libraries must be located on the computer running the application. Depending on the operating system, the process of locating linked libraries varies. I will explain in more detail how to deal with this when you encounter it in the book.

Digital Audio Workstation (DAW)

A DAW is like an IDE for musicians. It is a collection of tools that work together to allow for the recording and production of music. DAWs support plugins – external software components that can be loaded and used for sound synthesis, effects processing and MIDI processing.

Plugin

A plugin is a software component that works inside a larger program. We will create plugins that work inside DAWs. The plugins we create will include machine learning capabilities.

Plugin host

A plugin host is any software that can load and use external plugins. DAWs are plugin hosts, but we will also use a simpler plugin host in some of our examples to help test our software and to allow our software to host and control plugins.

2.1.2 Computer hardware

Now you have learned about the development environment's main components, it is time for a brief discussion of computer hardware. I have used several computer systems to write this book and develop the cross-platform software within it. These include a Windows 10 Intel i7 machine with a discrete Nvidia Geforce 2070 graphics processor (GPU), an 'Apple Silicon' Mac with an M1 chip, an Intel Mac with an Intel i5 CPU, both running macOS and an Ubuntu Linux Intel machine with integrated Intel GPU. Do not worry if you cannot access as many different machines. You only need one machine to work through the book. But you will need access to those systems if you want to build versions of the programs for Windows, macOS and/ or Linux. You will also need administrative rights to install software on your development machine(s) unless it has already been installed for you.

GPUs, training and inference

If you have attempted to use a machine learning model before, you may have thought about the role of the GPU. For example, it is common for artificial intelligence researchers to publish executable Python notebooks with their research papers, and you may have experimented with one of these. Often notebooks are set up to run on cloud computing services such as Google colab[2]. When you run the notebook in the cloud, you will find various GPU options available in the user interface.

So why do machine learning systems use GPUs? This is an interesting question because using GPUs for machine learning was one of the breakthrough ideas that enabled the deep learning revolution to begin in the late noughties. Geoffrey Hinton, sometimes called the 'AI Godfather', presumably in the spiritual mentor as opposed to crime-boss sense was one of the pioneers in the use of GPUs for deep learning. As deep learning pioneer LeCun puts it, GPUs "were convenient to program and allowed researchers to train networks 10 or 20 times faster"[23]. In the same paper, the researchers mention that speech recognition was one of the first applications. So audio signal processing was a core application for deep learning from its inception.

So machine learning, particularly deep learning, runs faster on GPUs. Why is that, given that GPUs are actually designed to process graphics? The original purpose of a GPU is to compute parts of the graphics pipeline. There are several parts of the graphics pipeline requiring the multiplication of matrices. For example, computing lighting effects, applying textures and updating the positions of moving objects. GPUs have been designed to compute many matrix multiplications in parallel. Unlike CPUs, GPUs have many cores – the Geforce 2070 in my laptop has 2,304 cores, whereas the CPU has eight. GPU cores are simpler and more specialised than CPU cores. Essential parts of machine learning algorithms can be boiled down to many matrix multiplications, and GPUs excel at these.

At this point, you may worry that you need a powerful GPU to work through the practical activities in the book, but that is not necessarily the case. Firstly, not all machine learning algorithms in the book require heavy computation. As well as that, the machine learning algorithms in the book that do require heavy computation only need it in their training phase. Training is when the algorithm learns. Inference is when we use the model to carry out a task. Remember that one aim of the book is to build applications that music and audio professionals can access – well that means we need to make sure the inference part runs fast enough on a CPU. Still, it will be useful to have access to a GPU sometimes – so check out the information later about Google Colabs for cloud-based GPU use. I will also provide pre-trained models where appropriate.

---

[2]https://colab.google.com/

2.2 Setting up for C++ development

FIGURE 2.3
Setting up your development environment involves complex machinery and lots of steps.

At this point, you should be familiar with the purposes of the critical components you will encounter in your development environment. Now it is time to put that knowledge into practice by setting up the components on your machine(s) and attempting to build and run some programs.

I have taught programming for many years in several languages. My experience has been that getting that first program compiled and running can be quite a frustrating first step. The reason is that the programmer needs to engage with many complex tools and concepts, even though they have not written a line of code yet. This problem is especially true when working in C++ with an IDE. Even worse, we are working with two different libraries, JUCE and libtorch and potentially, cross-platform!

So, I want you, the reader, to know that the next few steps can go wrong and be quite frustrating, but stick with it and be prepared to search online for any error messages you encounter. Eventually, you will have all the tools humming away nicely. At the end of this section, I have created a list of common issues people encounter with the development environment setup process along with possible solutions.

If you already have an IDE installed and can compile and run C++ programs, it is still worth reading through this section, as a few special steps are required for the projects in this book. For example, you might need to become more familiar with JUCE, CMake or libtorch and how they work together.

As for all software setup and configuration instructions in the book, minor variations between operating systems and software versions might cause glitches, and things might break when new software versions are released. If you encounter any glitches, the best thing to do is to post an issue on the GitHub page for the course materials.

We will start with the essential tools that allow us to build software: an IDE and the tools for C++ development. What you do here depends on the operating system(s) and hardware you have available to you. Instructions are here for all three operating systems supported by the book. You can skip the sections for the operating systems you are not using.

FIGURE 2.4
Creating, building and running a C++ console program in Visual Studio.

2.3 Windows: install Visual Studio Community Edition

If you are developing for Windows, install Visual Studio Community Edition. You can move to the slightly simpler Visual Studio Code later, but installing Visual Studio Community Edition first will make it easier to get up and running with C++. I have tested the code for the book in Visual Studio 2022 Community Edition, which is free. Your first job is to search for 'Visual Studio Community Edition download' in your favourite search engine, then download and install it. You will need to select and install extra packages during the setup process. I have installed the C++-related packages shown in figure 2.5 on my system, and I recommend that you do the same. Once everything is installed, you can launch Visual Studio and create, build and run a new C++ console project. Figure 2.4 contains screenshots showing those three stages.

FIGURE 2.5
C++-related packages that you should install.

2.4 macOS: install Xcode

I recommend that you use Xcode for macOS development. As for Windows, this is probably the easiest way to get up and running with C++ development. You can install Visual Studio Code later if you want a consistent development environment across platforms. The simplest way to install Xcode is from the Apple App Store, but the large download from the App Store can be unreliable. An alternative method is registering and logging into the Apple Developer portal and accessing

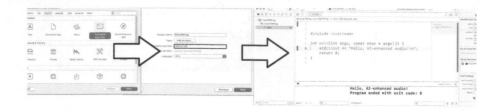

FIGURE 2.6
Creating, building and running a C++ console program in Xcode.

the Xcode download. This download is more reliable, and you can also select older versions of Xcode in case you have an older version of macOS. Once you have installed Xcode, you must install the command line tools. You can do this by running the following command in the Terminal app:

```
sudo xcode-select --install
```

Once all that is ready, you can create, build and run a C++ command line program in Xcode. Figure 2.6 contains screenshots showing those three stages.

2.5 GNU/Linux: install Visual Studio Code

If you work under GNU/Linux, which I will henceforth shorten to Linux, you might already have a preferred program for editing source code. I have worked for many years in Emacs (including writing my PhD thesis there); I also use Nano for command line editing and Gedit for a simple graphical editor. Recently, I have moved to Visual Studio Code (VSCode) for most programming activities. VSCode is open-source and has excellent extensions for working in C++ and CMake. I am writing this book in VSCode using James Yu's excellent LaTeX Workshop extension. You can download VSCode from the Microsoft website or find it in the application repository for your Linux distribution. There are further instructions on configuring Visual Studio Code in the next section. You do not have to use this IDE – feel free to use Vi or Emacs or whatever you prefer. The projects in the book only require CMake amd the regular build tools to build, so you can edit the source code however you like.

Regardless of your IDE choice, you will likely need to use your package manager to install the C++ build tools and CMake. For example, on Ubuntu, I run the following command:

```
sudo apt install build-essentials cmake
```

Let's see if you can create, compile and run a C++ program in Linux. Put this into your text editor of choice:

```
#include <iostream>
int main(){
    std::cout << "Hello, AI-enhanced audio!" << std::endl;
}
```

Save it as 'main.cpp' and run this command to compile and link it:

```
g++ main.cpp -o myprogram
```

Finally, run it with this command:

```
./myprogram
```

If everything is set up correctly, you should see a program called myprogram appearing, and it should print out "Hello, AI-enhanced audio!" when you run it.

2.6   All platforms: Visual Studio Code

To use VSCode, download the installer for your operating system from the Microsoft website. Then you will need to add some extensions. On my VSCode setup, I have the following extensions installed: C/C++, C/C++ Extension pack, C/C++ Themes and CMake Tools, all from Microsoft. You can install extensions by clicking the cog in the VSCode sidebar and selecting 'Extensions' from the menu that pops up. As mentioned earlier, I recommend first installing an IDE and/ or the build tools you need.

2.7   Progress check: build and run C++ hello world

At this point, you should be able to build and run a simple C++ hello world program in your IDE. If you cannot do that, you need to keep working on your setup until you can, then proceed to the next step.

FIGURE 2.7

My development setup showing an M1 Mac running macOS hidden on the left, a ThinkPad running Ubuntu 22.04 in front of the monitor and a Gigabyte Aero running Windows 11 on the right.

2.8 My development setup

What kind of setup do I use for cross-platform AI-enhanced audio development? My setup includes an M1 Mac Mini for macOS builds, a Gigabyte Aero with Geforce 2070 GPU for Windows/ Linux and accelerated GPU training and a Thinkpad X1 for general Linux work. I use RME audio interfaces and Genelec speakers. I generally work in Visual Studio Code, which provides me with a consistent experience on all three operating systems, and the CMake support in VSCode is quite advanced. I also have Xcode and Visual Studio Community 2022 to test my builds. I have found that musicians often have ancient versions of macOS, so I also keep an Intel Mac around with an older macOS version for testing.

3

Installing JUCE

This chapter will teach you about the JUCE library and associated tools. JUCE allows you to create C++ software projects which can be exported for multiple platforms from the same code-base. I will talk you through your first plugin build using Projucer and an IDE, covering Windows, macOS and Linux. I will also show you how you can test your plugins using the AudioPluginHost tool, a plugin host that comes with the JUCE library. At the end of the chapter, you will be building, running and testing plugins on your machine.

3.1 Why JUCE?

Now you have a working C++ development environment, you are ready to add the next layer: JUCE. JUCE, standing for Jules' Utility Class Extensions, was developed initially by Julian Storer. JUCE will allow you to build cross-platform native audio plugins that integrate into standard audio software. JUCE is not the only way to develop cross-platform audio plugins. There are other similar libraries, such as Oli Larkin's iPlug2, or you could even directly use a plugin development API, such as Steinberg's VST3. I have chosen to use JUCE for the following reasons: it provides a very consistent experience on different platforms; it includes a set of cross-platform user interface components; it is compatible with the CMake build tool, and it can export plugins in several formats, such as VST3, Audio Unit and so on. JUCE also provides components to build applications and plugins that can host other plugins.

Some people consider JUCE to be an opinionated library which forces you to work in a particular way. This is probably true, but it is reasonably consistent and my students generally get up to speed with writing applications in JUCE quite quickly. However, you should be able to port the concepts from the book to other plugin APIs if you want to. It all comes down to user interface widgets and audio buffers – it just depends on how much time you want to spend making them work, how much engineering freedom you want and what your licensing requirements are.

3.2 Installing JUCE

To install JUCE go to their website[1] and download it for your platform. Alternatively, you can visit the JUCE GitHub releases page[2] and download from there. You will end up with a folder containing several subfolders. There is a 'docs' folder containing documentation for the JUCE API (the API is the collection of ready-made components/ classes that come with JUCE). There is an examples folder containing example JUCE projects. There is an extras folder containing additional projects, some of which are examples, some of which are utilities. One such utility is the AudioPluginHost which I will show you how to use later to test your plugins.

3.3 Projucer

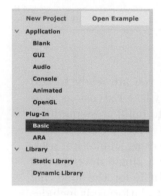

FIGURE 3.1
The available application types for Projucer.

Projucer is a program that comes with the JUCE distribution. Projucer is a central hub for managing your JUCE projects. It can generate different types of projects for different OS and IDE combinations. Figure 3.1 shows the kinds of applications for which Projucer can generate templates. The most interesting one for our purposes is Basic Plugin.

Go ahead – run Projucer and create a new 'Basic Plugin' project. You will be prompted to name it and save it somewhere. You should end up on the Projucer project screen, as shown in figure 3.2. If you click on the Modules tab on the left, you can check if your JUCE system is configured correctly – the modules should not be highlighted in red. If they are red, check that the 'Global paths' options are set correctly from the file menu to point to where you have unzipped the JUCE folder. Clicking on the Exporters tab lets you add new exporters by clicking the add button. Make sure there is an exporter for your system.

Now you are ready to save the project for use in your IDE. Xcode and Visual Studio users can select the exporter they want from the dropdown at the top of the Projucer menu and click the icon to the right of it. This will launch the project in

---

[1] https://www.juce.com
[2] https://github.com/juce-framework/JUCE/releases

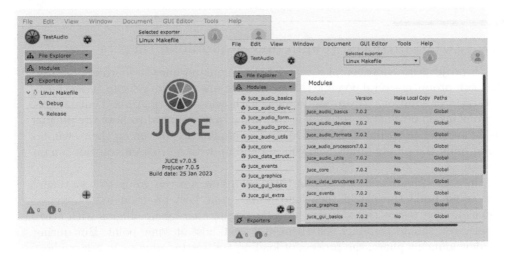

FIGURE 3.2
Projucer project view. The exporter panel is exposed on the left, module configuration panel is on the right. Note that my modules are all set to Global.

your IDE. Linux users will need to execute the Makefile from the terminal. More on that shortly. Saving the project should generate a folder hierarchy like the one shown below.

```
|-- Builds
|   |-- LinuxMakefile
|   |-- macOSX
|   '-- VisualStudio2022
|-- JuceLibraryCode
|-- MyNewPlugin.jucer
'-- Source
    |-- PluginEditor.cpp
    |-- PluginEditor.h
    |-- PluginProcessor.cpp
    '-- PluginProcessor.h
```

The Builds folder contains the IDE project files. The Source folder contains the source code for your project that you will end up editing. The .jucer file is the Projucer configuration file.

3.4   JUCE projects: plugin and Standalone targets

JUCE projects are more complex than a basic command-line C++ project because they can produce several possible things, not just one. We can refer to these things as targets or solutions. If you are familiar with using VST or Audio Unit plugins on your music production setup, you will know that you need to host the plugins inside another program, such as a DAW. The plugins do not run on their own. So if you build the plugin target, you will not end up with a program you can run on its own.

Since the plugins do not run on their own, how are you going to test them? You could build the plugin then load it into your host program (e.g. Logic or Cubase), then test it there. If you are planning to share your plugins with other people, you should certainly test them in a 'real' host at some point. But during the development phase, the process of loading and testing plugins in a large host application can be quite slow and fiddly.

Luckily, JUCE plugin projects have a handy target called the Standalone target. The Standalone target wraps your plugin into an executable application that you can run independently of a plugin host. It is much quicker to carry out the build and test cycle with the Standalone target.

3.5   Build and run a JUCE Basic Plugin on Windows

Now you are going to find out how to build JUCE plugin projects on different platforms. Firstly, Windows and Visual Studio. To build and run the project in Visual Studio, click the Visual Studio icon in Projucer, or go into the Builds folder and open the project from there. If you have multiple versions of Visual Studio installed on your system, check that the project opens in the correct version of Visual Studio. In my case, that is Visual Studio Community 2022. Once the project loads, select the Standalone project from the solution area and then select 'Build Solution' from the build menu. It might print a few warnings and if everything goes well, no errors during the build. If the build fails, check common errors in section 7. Finally, select 'Run without debugging' from the Debug menu, and your plugin should launch in Standalone mode.

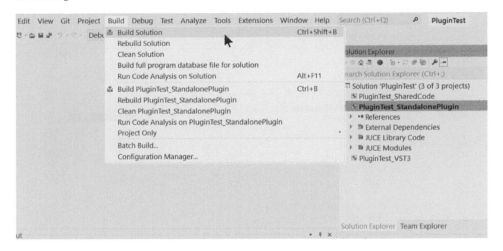

FIGURE 3.3
Building the Standalone solution in Visual Studio Community 2022.

3.5.1 Displaying JUCE debug output in Windows

There is one more step you need to take on Windows. JUCE provides a function (really a macro) called DBG, which allows you to print text to the console. By default, Visual Studio will not show this text output anywhere. To see the output, open settings from the tools menu and search for 'immediate'. As shown in figure 3.4, check the option to redirect text to the 'Immediate Window', then show the Immediate Window by searching or using the keyboard shortcut. Figure 3.4 illustrates a Standalone plugin running in Visual Studio debug mode with the Immediate Window showing text output.

3.6 Build and run a JUCE Basic Plugin on macOS

To build and run the project in Xcode, click the Xcode icon in Projucer to launch Xcode, select the Standalone option as shown in figure 3.5, then press the play button on the interface. If everything is set up correctly, it should report that the build was successful and launch the application. Output from the DBG macro should appear in Xcode's built-in console.

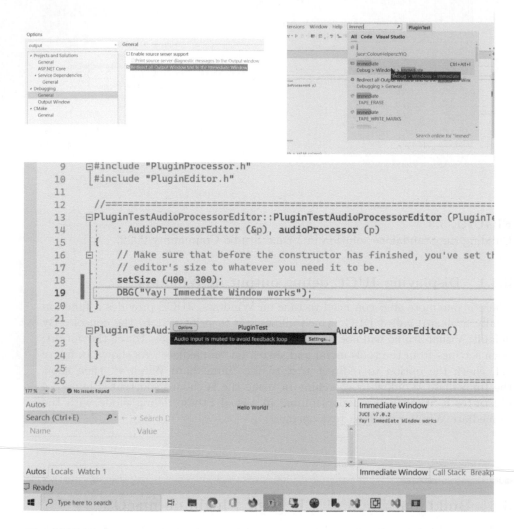

FIGURE 3.4

Enabling console output for a JUCE project in Visual Studio/ Windows. At the top: redirect text output to the immediate window and open the immediate window. At the bottom: a program running with DBG output showing in the immediate window.

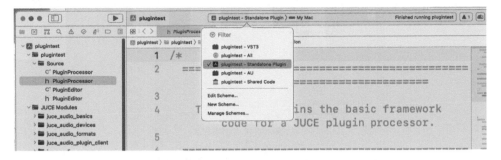

FIGURE 3.5
Running a JUCE plugin project in Xcode – make sure you select Standalone.

3.7 Build and run a JUCE Basic Plugin on Linux

To build the project in Linux, create the project in Projucer, make sure you have added the Linux Makefile exporter and select 'Save Project' from the Projucer file menu. Open a terminal and cd into the location of your saved project. You should see a Builds folder that contains a LinuxMakefile folder. cd to the LinuxMakefile folder. Finally, run the command:

```
make -j 4 # increase 4 if you have lots of CPU cores
```

This will build your project and store it in a build folder. To run it:

```
./build/YourProjectName
```

If this seems a bit 'manual', do not worry – we will see a more powerful way to build and run JUCE projects on Linux once we install CMake.

3.8 Test a plugin using JUCE's AudioPluginHost

At this point, you should have successfully built and run a JUCE Basic Plugin on your system. If you still need to, please continue working on the setup until you have. To complete this system setup phase, we will load the Basic Plugin into a plugin host instead of running it in Standalone mode. The Standalone mode is provided as a 'test harness' so you can quickly run and test the plugin without the complexity of running it inside a DAW or other plugin host. JUCE has a sophisticated plugin host designed to help you test your plugins in real 'plugin' mode. We will now build and run that plugin host and import our test plugin.

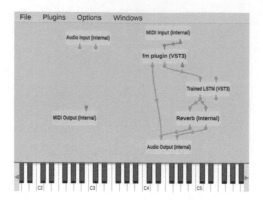

FIGURE 3.6
The JUCE AudioPluginHost application, which comes with the JUCE distribution. One of its built–in plugins, a sine synth, is wired to the MIDI input and the audio output.

The plugin host is part of the download for JUCE. It is located in the JUCE/extras/AudioPluginHost folder. You will find a .jucer file there that you can open with Projucer. Save the project from Projucer, then build and run it as usual. You should see an interface like figure 3.6. The AudioPluginHost application is a graph-based environment which allows you to insert plugins into a graph and then connect them to MIDI data inputs and outputs and audio inputs and outputs. The AudioPluginHost application has some ready-made plugins that you can wire into the graph.

How can you wire your plugin to the graph? The first thing to know is that audio plugins are installed to standard locations on your system. When a plugin host such as Reaper or Cubase starts up, it scans these locations for new or updated plugins. Then the host knows which plugins are available. The AudioPluginHost can also scan these folders to find plugins. The standard locations for VST3 plugins are:

Windows:

```
C:\Program Files\Common Files\VST3
and
<home directory>\AppData\Local Programs\Common\VST3
```

Mac:

```
/Library/Audio/Plugins/VST3
and
<home directory>/Library/Audio/Plugins/VST3
```

Linux:

```
/usr/lib/vst3/
and
<home directory>/.vst3/
```

So for plugin hosts to see your plugins, they should be in one of those folders. Luckily the JUCE plugin project includes a step in the build process wherein plugins are copied to the appropriate locations. You can check this is enabled in Projucer by clicking on the exporter, selecting Debug or Release and looking for 'Enable Plugin Copy Step', which should be enabled.

Rebuild your project with the plugin copy step enabled, then run the AudioPluginHost. Go to the options menu in the plugin host and edit the list of available plugins. A list of available plugins should appear, as shown in figure 3.7. You can then trigger a scan for new plugins, and your plugin should appear in the list. It will probably appear under the company name

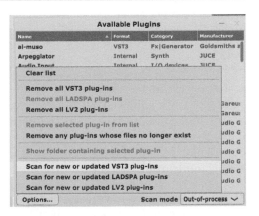

FIGURE 3.7
The list of available plugins in the JUCE AudioPluginHost app. I have clicked the options menu which is showing its 'scan for new or updated ...' function.

'yourcompany' as that is the default in a new Projucer project. Try inserting your plugin into the graph. It does not do anything yet, so we'll not hear anything. Do you have another plugin host on your system? If so, try loading the plugin into that too.

3.9   Progress check: build and run a JUCE plugin

At this point, you should be able to run the Projucer application, use it to generate a plugin project, and then build and run the plugin in standalone mode from your IDE. You should also be able to load your plugin in pure plugin mode into a plugin host. You should have tried this with the JUCE AudioPluginHost application and possibly another host.

4

Installing and using CMake

Now that everything is working, we are going to go back a step and break everything again(!). We will swap out the Projucer application for an alternative project generator tool called CMake. This chapter shows you how to get up and running with JUCE and CMake. CMake is a more general-purpose tool than Projucer, and it is particularly good at integrating libraries into projects if they also use CMake. The machine learning library libtorch and the neural network library RTNeural are examples of libraries we'll use later that support CMake. You may wonder why we bothered with Projucer instead of going straight for CMake. Three reasons: 1) I wanted to lead you in gently, 2) Projucer is helpful for running and building example projects from JUCE, and it provides a more straightforward method to verify that you have an IDE with C++ capabilities and JUCE 3) Putting CMake in the mix too early would provide too many opportunities for frustrating configuration problems. Don't worry – setting up CMake should be straightforward if you are here and you have a working build system.

4.1 Install CMake on Windows

To install CMake on Windows, follow the instructions on the CMake website[1]. Once installed, you should be able to fire up your favourite shell program, Powershell, in my case, and type CMake at the command prompt. Figure 4.1 shows the results of running CMake in Windows Powershell.

---

[1] https://cmake.org

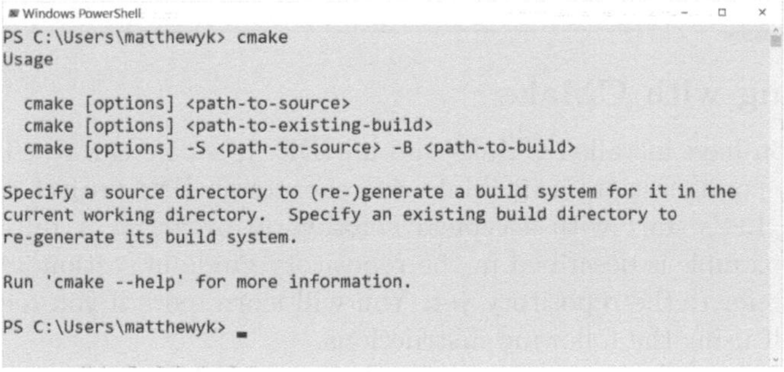

FIGURE 4.1
CMake running in the Windows Powershell.

4.2 Install CMake on macOS

I recommend installing Homebrew for macOS as this makes it easy to install CMake and allows for flexible Python setups, amongst other things. Please follow the instructions on the homebrew website to install homebrew[2]. Once you have installed homebrew, run the following command in the Terminal app to install CMake:

```
brew install cmake
```

4.3 Install CMake on Linux

Installing CMake on Linux can be achieved using the package manager. For example, if you use Debian/ Ubuntu:

```
sudo apt install cmake
```

---

[2]https://brew.sh/

4.4 Working with CMake

I will assume you have installed CMake and an IDE. If so, we can test if you can create a CMake configuration, use CMake to generate an IDE project then build it in your IDE. Let's start with a typical hello world program. A fully working version of this example is described in the repository guide in section 39.2.1. You do not have to refer to the repository yet. You will learn more if you recreate the example yourself using the following instructions.

Start by creating a folder for your project. In that folder, create a file called CMakeLists.txt (noting the case of the letters). Put the following text into the CMakeLists.txt file:

```
cmake_minimum_required(VERSION 3.0 FATAL_ERROR)
project(hello-cmake)
add_executable(hello-cmake src/main.cpp)
```

That text contained a set of CMake commands. The command `cmake _minimum_required` instructs CMake to fail if it is not at least version 3.0. The `project` command names the project. The `add_executable` command adds an executable to the project called `hello-cmake`, which is to be built from the file `src/main.cpp`.

The next step is creating the `src/main.cpp` file. Add a subfolder to your project's folder called `src` and create a file named `main.cpp` in there. Put a classic hello world program in `main.cpp`. Mine looks like this, but feel free to put a more enthusiastic message in there:

```
#include <iostream>
int main(){
  std::cout << "Hello CMake" << std::endl;
}
```

Your project folder should now have a file hierarchy like this:

```
.
|-- CMakeLists.txt
'-- src
    '-- main.cpp
```

Now we are ready to use CMake to generate an IDE project. Fire up your preferred shell—for me, Terminal.app on macOS, Powershell.exe on Windows or the regular Terminal program on Ubuntu. Change the directory inside the command shell to the project folder you just created. That will probably involve the `cd` command. Once in the directory, run this CMake command:

```
cmake -G
```

This will list all the available project generators that CMake can use. Remember that the idea is to configure the build using CMake then to generate projects for the various IDEs. This is conceptually similar to how we used Projucer. We are swapping out Projucer for CMake. The command output should also show you which generator is the default. The default generator on my Windows 10 system is 'Visual Studio 17 2022'. It also lists 'Visual Studio 16 2019' as an option. The default on my macOS and Ubuntu systems is 'Unix Makefiles', but the Mac also lists 'Xcode'. To generate the project for a specific target, specify that target using the G option:

```
# Windows/ Visual Studio version:
cmake -G "Visual Studio 17 2022" -B build .
# macOS/ Xcode version
cmake -G "Xcode" -B build .
# Linux/ Makefile version
cmake -G "Unix Makefiles -B build .
```

The B option tells CMake to create a folder called `build` and to output its generated project into that folder. The full stop at the end ('.') tells it that the CMake configuration file is in the current folder. So this command should create a folder called `build` containing your IDE project.

4.5   CMake command-line build: any platform

It is time to actually build the target. The simplest way to build the CMake project is by running a CMake command like this:

```
cmake -B build .
cmake --build build
```

Those commands will generate a folder called 'build' and then build the project into that folder. Whilst that is the simplest possible command sequence to build a CMake project, CMake actually generates multiple forms of the build. For the debug build (no compiler optimisations, debug flags set, assertions asserted):

```
cmake --build build --config Debug
```

Now to carry out the release build (compiler optimisations on, no debug flags, assertions not asserted):

```
cmake --build build --config Release
```

Note that for this simple project, running the second build command will wipe out the previously built debug-mode executable.

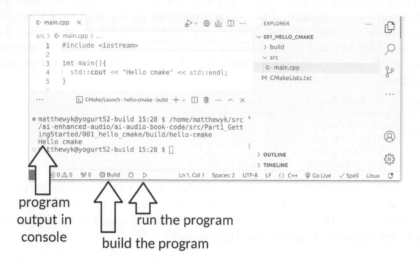

FIGURE 4.2
A CMake project viewed in VSCode.

4.6   Build in VSCode: any platform

Using the cross-platform standard VSCode setup described in section 2.6, you can easily build the hello-cmake CMake project. First, if you have been experimenting with CMake on the command line outside an IDE, you might have a build folder in the same folder as your CMakeLists.txt file. Delete that folder, as it can cause confusion when VSCode takes over managing the build folder.

Next, launch VSCode and select 'open folder' from the file menu. Browse to the folder containing the CMakeLists.txt file and open it. You need to install the Microsoft 'CMake Tools' extension. You might be prompted to do this when you open the CMakeLists.txt file, or you might need to go into your VSCode settings/extensions to find the extension and install it. This extension will allow VSCode to parse the CMakeLists.txt file. When you open a folder containing a CMakeLists.txt file in VSCode, VSCode should automatically detect the file and configure the build. Before doing this, VSCode might ask you to select a build kit. You will see different options here, depending on your operating system. To see which kits are available, open the VSCode Command Palette from the View menu and run the 'Cmake: Scan for Kits' command. Here is some of the output from that command on my Windows 10 machine:

```
[kit] Found Kit: Visual Studio Community 2022 Release - x86
[kit] Found Kit: Visual Studio Community 2022 Release - x86_amd64
[kit] Found Kit: Visual Studio Community 2022 Release - amd64_x86
[kit] Found Kit: Visual Studio Community 2022 Release - amd64
...
[kit] Found Kit: Clang 15.0.1 (MSVC CLI) for MSVC 17.5.33414.496
 (Visual Studio Community 2022 Release - x86)
[kit] Found Kit: Clang 15.0.1 (GNU CLI) for MSVC 17.5.33414.496
 (Visual Studio Community 2022 Release - x86)
...
```

What is the difference between the Visual Studio 2022 options: x86, x86_amd64, amd64_x86 and amd64? The first part is the CPU architecture for the machine used to build the binary, and the second part is the CPU architecture for the machine(s) that will run the binary. If there is only one part (amd64), the builder and target are the same. amd64 is a 64-bit architecture, and x86 is a 32-bit architecture. If you are on a 64-bit machine running 64-bit Windows, choosing amd64 is fine.

What about macOS? Here is the output I see on my m1 macOS machine (only one kit was found):

```
[kit] Found Kit: Clang 14.0.0 arm64-apple-darwin22.3.0
```

Here is some of the output I see on my Ubuntu Linux machine. I had to open the output area from the View menu and then select CMake/ build from the list of possible sources to see the list:

```
...
[kit] Found Kit: GCC 11.3.0 x86_64-linux-gnu
[kit] Found Kit: Clang 14.0.0 x86_64-pc-linux-gnu
[kit] Found Kit: Clang 11.1.0 x86_64-pc-linux-gnu
---
```

The naming convention for build and target machine described for Windows does not apply to Linux – x86_64-linux-gnu builds on 64-bit and targets 64-bit.

Once you have selected your build kit, click the 'build' button at the bottom of VSCode to compile and link, then the play icon to run the program. You should see the output of the hello-cmake program in VSCode's built-in Terminal.

4.7   Build in Visual Studio: Windows

If you are working with Microsoft Visual Studio on Windows, there are two ways to work with CMake projects: using CMake to generate the project or using Visual

run the program   available targets

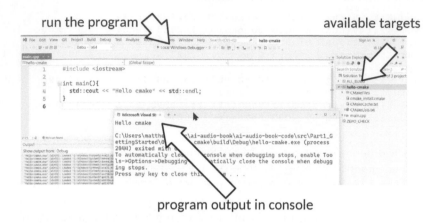

program output in console

FIGURE 4.3
A CMake project viewed in Visual Studio Community 2022.

Studio's built-in CMake support. I recommend the first option as I have encountered issues with the second approach, though it does produce neater-looking JUCE projects. The following instructions relate to projects generated from the CMake command.

After running the CMake command with the -G option described above, you should see a file called **hello-cmake.sln** in the **build** folder. Find this file in Windows Explorer – the easy way is to run the command **explorer.exe**. in Powershell when in the **build** folder. Then open the **.sln** file in Visual Studio. A **.sln** file is a VS solution file, a collection of 'products'. Each product within the solution has a separate Visual Studio project file. Figure 4.3 shows the hell-cmake project in Visual Studio, highlighting the list of targets, console and run button. The project will start with ALL_BUILD as the target for the play button, so you should right click on hello-cmake in the solution explorer and select the 'Set as Startup Project' option. Clicking the play button now builds and runs hello-cmake.

run the program   available targets

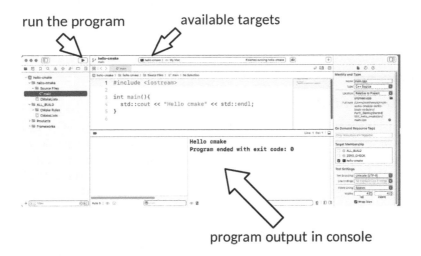

program output in console

FIGURE 4.4
A CMake project viewed in Xcode.

4.8   Build in Xcode: macOS

Run this command in the Terminal application in the folder containing the CMake-Lists.txt file:

```
cmake -G "Xcode" -B build .
```

That will generate an Xcode project in the build folder which you can open with Xcode. After opening the project in Xcode, hit the play button, and it will build and run the program. Figure 4.4 illustrates what you will see when you open a CMake project in Xcode.

4.9   Building the project for GNU/Linux

I recommend using the Standard VSCode setup for Linux, but I appreciate that Linux users might prefer their own setup. This is especially true for those who bitterly remember Steve Ballmer referring to Linux as a virus and do not wish to use any Microsoft tools. In that latter case, please refer to the command line instructions above, or read the documentation for the CMake support in your preferred IDE.

4.10   CMake and JUCE

Now that you have installed CMake and can build C++ projects, it is time to return to JUCE and convert the plugin project to CMake. I am instructing you to use CMake instead of Projucer because it is easier to integrate libtorch and other CMake-compatible libraries into a CMake project than the standard Projucer project. This project is described in the repository guide, section 39.2.2.

4.10.1   File structure for a CMake JUCE plugin project

The first step is to set up the file structure for your plugin starter project. Consider the structure below:

```
.
|-- CMakeLists.txt
|-- README.md
'-- src
    |-- PluginEditor.cpp
    |-- PluginEditor.h
    |-- PluginProcessor.cpp
    '-- PluginProcessor.h
```

You can obtain some template PlugEditor and PluginProcessor files by generating a Basic Plugin Project in Projucer and copying over the contents of the Source folder. Or you could start from the CMake version of the project described in the repository guide section 39.2.2.

The CMakeLists.txt file should contain the following lines. Do not worry too much about what everything here is doing yet. Several of these commands are custom macros provided by the JUCE library's CMake API. You can find more information about the JUCE CMake API on the github repository[3]. These macros automatically become available to CMake because the JUCE library gets added to your project with the add_subdirectory command. I will explain the essential parts which you need to change below and the others when we need them. Put the following text into your CMakeLists.txt file (or access the example from the repository):

---

[3]https://github.com/juce-framework/JUCE/blob/master/docs/CMake API.md

```
1   cmake_minimum_required(VERSION 3.15)
2   project(minimal_plugin VERSION 0.0.1)
3   # where is your JUCE folder? ../../JUCE for me
4   add_subdirectory(../../JUCE ./JUCE)
5   juce_add_plugin(minimal_plugin
6       COMPANY_NAME Yee-King
7       # set to false if you want audio input
8       IS_SYNTH TRUE
9       NEEDS_MIDI_INPUT TRUE
10      IS_MIDI_EFFECT FALSE
11      NEEDS_MIDI_OUTPUT TRUE
12      COPY_PLUGIN_AFTER_BUILD TRUE
13      PLUGIN_MANUFACTURER_CODE Yeek
14      # should change for each plugin
15      PLUGIN_CODE Abc1
16      FORMATS AU VST3 Standalone
17      # should change for each plugin
18      PRODUCT_NAME "minimal_plugin")
19
20  juce_generate_juce_header(minimal_plugin)
21
22  target_sources(minimal_plugin
23      PRIVATE
24      src/PluginEditor.cpp
25      src/PluginProcessor.cpp)
26
27  target_compile_definitions(minimal_plugin
28      PUBLIC #
29          JUCE_ALSA=1
30          JUCE_DIRECTSOUND=1
31          JUCE_DISABLE_CAUTIOUS_PARAMETER_ID_CHECKING=1
32          JUCE_USE_OGGVORBIS=1
33          JUCE_WEB_BROWSER=0
34          JUCE_USE_CURL=0
35          JUCE_VST3_CAN_REPLACE_VST2=0)
36
37  target_link_libraries(minimal_plugin
38      PRIVATE
39          juce::juce_audio_utils
40      PUBLIC
41          juce::juce_recommended_config_flags
42          juce::juce_recommended_lto_flags
43          juce::juce_recommended_warning_flags)
```

Note that setting the COPY_PLUGIN_AFTER_BUILD property in the juce_add_plugin block to TRUE is equivalent to editing the Projucer settings to enable the copy plugin step.

You might add the header files to the target_sources command as well, if that is appropriate for your setup. I have seen people adding header files when working in Visual Studio Community as it makes them easier to find in the project.

4.10.2 Customise the CMakeLists.txt file

You must apply some customisations to make that CMakeLists.txt file work for you. Firstly, CMake needs to know where JUCE is on your system to build your project and access the CMake macros. Update the add_subdirectory command on line 3 to point at your JUCE directory. The command is written like this:

```
add_subdirectory(../../JUCE ./JUCE)
```

The first '../../JUCE' tells CMake where JUCE can be found. I keep JUCE two folders up from my current project directory, so I specify '../../'. You can put any path you like there. The second part, './JUCE', defines where JUCE will be copied to in the build folder. You can leave that as it is.

Next, the project title. Everywhere you see 'minimal_plugin' in the CMake-Lists.txt file, replace it with your chosen project title. This can be any string compatible with CMake's project naming conventions. You can use find and replace in your code editor to do it. I found the following commands need to be edited to implement a new project name:

```
project(minimal_plugin -> your-plugin
juce_add_plugin(minimal_plugin -> your-plugin
juce_generate_juce_header(minimal_plugin) -> etc.
target_sources(minimal_plugin
target_compile_definitions(minimal_plugin
target_link_libraries(minimal_plugin
```

Now you need to change the properties the plugin will report to your host. These are the company name and plugin identifiers your DAW displays when you search for plugins. In the section 'juce_add_plugin', change the following properties:

```
COMPANY_NAME Yee-King  # change to your company name
PLUGIN_MANUFACTURER_CODE Yeek # make one up!
PLUGIN_CODE Abc1 # each of your plugins needs a unique code
PRODUCT_NAME "minimal_plugin" # make one up!
```

4.10.3 Recap of JUCE, CMake and plugins

To recap, here is what you need to build a plugin using CMake:

1. A CMakeLists.txt file containing the text specified above, edited to suit your system and names

2. A src folder containing the PluginProcessor and PluginEditor cpp and header files, the same ones generated by Projucer when you create a Basic Plugin project

3. That's it!

4.10.4 Build and run your JUCE plugin

Hopefully, you have prepared the files as described above. As for the simple hello-cmake project we saw before, you can run CMake with the -G and -B options to generate projects for your IDE of choice. Or you can load the folder containing the CMakeLists.txt file into Visual Studio Code, and it should figure everything out for you and get ready to build the project. Or you can carry out the build entirely on the command line as described above. Common build and setup problems are discussed in section 7. You can try running the Standalone target if the project builds without errors. This is the same procedure you saw in section 3.4 when using the Projucer project. Remember you can refer to the example projects in the source code repository if you get stuck.

4.11 Progress check

At this point, you should be up and running with CMake. You should understand the purpose of a CMakeLists.txt and be able to create a simple version of that file. You should be able to generate an IDE project from a simple CMake project using CMake commands. You can load that project into your IDE, be it Visual Studio, XCode, Visual Studio Code or something else! You should also know how to work with CMake JUCE projects and realise that CMake replaces the need for Projucer.

5

Set up libtorch

The chapter explains how to install and use the libtorch machine learning library. You will start by examining a minimal example of a libtorch C++ program. Then you will dig into the libtorch installation process and see how to integrate libtorch into a simple CMake build. Following that, you will see how to combine libtorch and JUCE in a single project. At the end of the chapter, you will be ready to experiment with libtorch and neural networks in C++. Please note that these instructions are correct and tested on Windows, macOS and Linux at the time of writing the book, but there are a few moving parts here that are subject to change. Please refer to the repo guide example 39.2.10 which will provide the latest and most correct CMake configuration.

5.1   What is libtorch?

Libtorch is an open-source machine learning library that allows you to create, train, and use neural network architectures. It is the native core of the popular PyTorch library and includes the TorchScript system, which makes it easy to import models created using PyTorch. With libtorch's C++ API, you can easily integrate its functionality into your C++ audio applications to access a comprehensive neural network design, training, and inference library. PyTorch, and therefore, libtorch is widely used by machine learning developers and researchers, making it a powerful tool for those looking to incorporate machine learning into their projects.

5.2   Simple libtorch example

Before you start the installation, I would like to whet your appetite by showing you a minimal example of a libtorch C++ program. Consider the following program:

```
1  #include <torch/torch.h>
2  #include <iostream>
3
4  int main() {
5    torch::Tensor tensor = torch::rand({2, 3});
6    std::cout << tensor << std::endl;
7    return 0;
8  }
```

This is an elementary program, but if you can build and run it, you are ready to use a state-of-the-art machine-learning library in your own native applications. If you have some experience working in C++, you should understand most of what is going on. For clarity, I will step through the program line-by-line and explain it. Firstly:

```
1  #include <torch/torch.h>
2  #include <iostream>
```

These lines include the header file for the libtorch library and the C++ standard library iostream header. Remember that in C++, header files contain declarations of functions, constants and such. They tell your IDE and compiler which functions are available. When you include a header, you are making the things declared in that header available for use in your program. So we are making the libtorch and iostream libraries available for use in our program.

Note that the included files are in angle brackets ($<$ ... $>$). That impacts where the IDE searches for these headers for its autocompletion and code-checking functionality and where the compiler searches when it pulls in the headers during the compile phase. Using angle brackets means they are located in a standard or, at least, specified place on your system. So there needs to be a folder somewhere called torch with a file called torch.h, and the IDE and compiler need to know where that folder is. We will use CMake to tell the IDE and compiler where to look later. Next, we have the main function:

```
1  int main() {
2    torch::Tensor tensor = torch::rand({2, 3});
3    std::cout << tensor << std::endl;
4    return 0;
5  }
```

The main function is the entry point for the execution of your program. In other words, it is the first function to automatically be called when your program runs. This main function starts by creating an object of type Tensor from the torch namespace. That is something declared in the torch.h header. We assign that Tensor object to a variable called tensor. Remember that objects in C++ contain data and functions relating to that data. That means our tensor object contains some data and some Tensor-related functions. *Tensors* are the core data structures in neural network-based machine learning, and we will encounter lots of them in this book.

We created the Tensor object using a function from the libtorch library called rand. rand takes an array as an argument, in this case, containing a 2 and a 3. I am assuming you know what an array is. We then use the cout function from the standard library to print out the contents of the tensor variable.

Here is the expected output of the program:

```
0.9660   0.2080   0.5723
0.6885   0.2689   0.9441
[ CPUFloatType{2,3} ]
```

This may look like a matrix to you – a two-dimensional, rectangular data structure. In fact, it is a matrix. You can think of tensors as matrices that are not limited to two dimensions. The simplest tensor is a single number or a scalar; the next simplest is a one-dimensional list of numbers or a vector; then a two-dimensional structure like a matrix and on to three dimensions and beyond. Great, but how do you get this program to build and run on your system? Keep reading.

5.3   Install libtorch

Now that you have seen the minimal libtorch program, it is time to install libtorch so you can run it (and hopefully some much more interesting ones) on your machine.

The first step is obtaining and installing the libtorch library, part of the PyTorch framework. How you do this depends on your operating system and hardware. The first step is to visit the downloads page for the PyTorch framework, which can be found at https://pytorch.org/. The download page will allow you to select the library version, the OS, the package (Conda, Pip, libtorch, Source), the language (Python or C++/Java) and the compute platform (Nvidia's CUDA accelerated GPU platform or CPU). Select the stable version of PyTorch (1.12.1 at the time of writing), your OS, LibTorch, C++/Java and CPU. The page should present you with a download link. If you have a choice, choose the cxx11 ABI version, not the pre-cxx11 ABI. Again, if you have a choice, choose the release build. The download can be quite large at 150+MB.

Eventually, you should have a zip file downloaded with a name similar to 'libtorch-cxx11-abi-shared-with-deps-2.1.1-cu118.zip' (Linux), 'libtorch-macos-2.1.1.zip' (Mac) or 'libtorch-win-shared-with-deps-debug-2.1.1-cu118.zip' (Windows). Windows users please note – if you are doing debug builds you need the debug version of libtorch; otherwise your program will crash in a confusing and silent manner. Version 2.1.1 was the latest version at the time of writing. Unzip this archive, and you should see a top-level libtorch folder containing various subfolders. Put the libtorch folder somewhere sensible for now. I put it in a folder called src in my user account home folder. Later we will need to tell CMake where that libtorch folder is to make those includes work. I will explain how to do that shortly.

5.3.1 Note for Apple Silicon users

If you are using a recent Apple machine with an Apple Silicon CPU (m1, m2 etc.), there are no native builds of libtorch available at the time of writing. I was able to create a native build myself based on instructions on this GitHub repository[1]. Essentially, follow the instructions here[2], except you add an additional CMake flag. The instructions are:

```
1 git clone -b main --recurse-submodule https://github.com/pytorch/
     pytorch.git
2 mkdir pytorch-build
3 cd pytorch-build
4 cmake -DBUILD_SHARED_LIBS:BOOL=ON -DCMAKE_BUILD_TYPE:STRING=Release -
     DUSE_MPS=ON  -DPYTHON_EXECUTABLE:PATH='which python3' -
     DCMAKE_INSTALL_PREFIX:PATH=../pytorch-install ../pytorch
5 cmake --build . --target install
```

Running those commands worked on my m1 mac. As elsewhere in this chapter, refer to the GitHub repository for the book for the latest information.

5.4 Configure the CMakeLists.txt file

Now we have libtorch 'installed', well, copied onto our system, we can start to build the CMake project for the program. The CMake commands described here are adapted from an example `CMakeLists.txt` file provided on the pytorch.org website in an article entitled 'Installing C++ Distributions of PyTorch'.

Create a new folder for your project. I call it 'minimal-libtorch', but you can call it what you like. Put the following in a `CMakeLists.txt` at the top level of your project folder:

```
1 cmake_minimum_required(VERSION 3.0 FATAL_ERROR)
2 project(minimal-libtorch)
3 set(CMAKE_PREFIX_PATH "../../src_resources/libtorch/libtorch")
4 find_package(Torch REQUIRED)
5 set(CMAKE_CXX_FLAGS "${CMAKE_CXX_FLAGS} ${TORCH_CXX_FLAGS}")
6 add_executable(minimal-libtorch src/main.cpp)
7 target_link_libraries(minimal-libtorch "${TORCH_LIBRARIES}")
8 set_property(TARGET minimal-libtorch PROPERTY CXX_STANDARD 17)
```

---

[1]https://github.com/mlverse/libtorch-mac-m1
[2]https://github.com/pytorch/pytorch/blob/main/docs/libtorch.rst

Compared to the hello-cmake `CMakeLists.txt` file from one of the earlier examples, there are a few extra commands. `set(CMAKE_PREFIX_PATH ..` uses the CMake `set` command to specify where cmake can find libtorch. You should set this to the folder where you unzipped libtorch earlier. On my Linux machine, the folder I point to contains the following top-level files and folders:

```
bin
build-hash
build-version
include
lib
Share
```

Next, the `find_package` command tells CMake to verify we have the Torch package installed. When you run CMake, it will look for libtorch in the standard places and any other places you added using the `set(CMAKE_PREFIX_PATH ..` command. `set(MAKE_CXX_FLAGS ...)` tells CMake to add the torch flags to the compile command. Finally, `target_link_libraries` tells CMake to link the program against the torch library.

5.5   Windows DLLs

If you are on Windows, you will need some additional CMake commands. Put this at the end of your `CMakeLists.txt` file:

```
# The following code block is suggested to be used on Windows.
# According to https://github.com/pytorch/pytorch/issues/25457,
# The DLLs need to be copied to avoid memory errors.
if (MSVC)
  file(GLOB TORCH_DLLS "${TORCH_INSTALL_PREFIX}/lib/*.dll")
  add_custom_command(TARGET minimal-libtorch
                     POST_BUILD
                     COMMAND ${CMAKE_COMMAND} -E copy_if_different
                     ${TORCH_DLLS}
                     $<TARGET_FILE_DIR:minimal-libtorch>)
endif (MSVC)
```

This instructs the build system to copy the libtorch shared libraries into the same folder as the compiled application (.exe). I recommend that you check the final version of the `CMakeLists.txt` file from the repo guide example 39.2.10 rather than typing it by hand as these instructions might have changed since the book was written.

For other platforms: the build you create on your machine should link to libtorch in the place it is located on your machine. That means that if you want the software to run on other people's machines, you will probably need to have libtorch in the same place. So you should probably look into creating an installer if you want to share your software with other people. Unfortunately the details of creating installers are beyond the scope of this book. Please refer to the GitHub repository where I will actively maintain information that readers of the book request such as this.

5.6 Create the main.cpp file

Now create a sub-folder called 'src' as before and put the code from the listing above into src/main.cpp. Here is that code again. Please type this out so your fingers get used to the libtorch syntax.

```
1  #include <torch/torch.h>
2  #include <iostream>
3
4  int main() {
5    torch::Tensor tensor = torch::rand({2, 3});
6    std::cout << tensor << std::endl;
7  }
```

As before, you should end up with a folder structure like this:

```
.
|-- CMakeLists.txt
'-- src
    '-- main.cpp
```

5.7 Generate the IDE project

Now you are ready to generate the IDE project from the CMakeLists.txt file. As you did for the hello-cmake project, open your terminal program and cd into your project folder. Then run the CMake command as usual. If you are working in Visual Studio Code, you can open the top-level folder containing the CMakeLists.txt file.

```
cmake -B build .
```

Add the -G option to select the desired IDE target if necessary.

5.8 Build and run

Now you should have either a Makefile project (Linux), an MSVC .sln project (Windows) or an Xcode project (macOS), or Visual Studio Code has read the CMakeLists file automatically. Open the project in your IDE as you did before and build. Then run the program. As before, here is the expected output of the program:

```
0.9660   0.2080   0.5723
0.6885   0.2689   0.9441
[ CPUFloatType{2,3} ]
```

5.9 Libtorch, CMake and JUCE

In this section, we will combine everything we have installed so far into a project which shows how you can run machine learning code inside a JUCE plugin. Don't get too excited yet – the machine learning code will not do much, but we will see how JUCE can interact with libtorch.

Start from the default CMake and JUCE basic plugin project. You should have created this earlier, but you can also find it described in the repository guide in section 39.2.2. Once you have a copy of the project ready, you need to add some lines to the standard JUCE CMakeLists.txt file such that it links the project against libtorch. You might refer back to your earlier libtorch project to ensure you have the correct settings for your system, but here are the general steps to make it work.

In CMakeLists.txt, tell CMake where libtorch is installed, instruct it to find libtorch and add the libtorch compile flags:

```
set(CMAKE_PREFIX_PATH "../../src_resources/libtorch/libtorch") #
    location of libtorch
find_package(Torch REQUIRED)
set(CMAKE_CXX_FLAGS "${CMAKE_CXX_FLAGS} ${TORCH_CXX_FLAGS}")
```

You can put that above the target_link_libraries command in CmakeLists.txt. Add a line to the target_link_libraries command as follows:

```
1  target_link_libraries(fm-torchknob
2  PRIVATE
3      juce::juce_audio_utils
4
5      #### add this line ###
6      "${TORCH_LIBRARIES}"
```

```
 7
 8  PUBLIC
 9      juce::juce_recommended_config_flags
10      juce::juce_recommended_lto_flags
11      juce::juce_recommended_warning_flags)
```

That instructs CMake to link your project against the libtorch library. Then if you are on Windows, you need to add this:

```
1  if (MSVC)
2  file(GLOB TORCH_DLLS "${TORCH_INSTALL_PREFIX}/lib/*.dll")
3  add_custom_command(TARGET fm-torchknob
4                     POST_BUILD
5                     COMMAND ${CMAKE_COMMAND} -E copy_if_different
6                     ${TORCH_DLLS}
7                     $<TARGET_FILE_DIR:fm-torchknob>)
8  endif (MSVC)
```

Ensure you set 'fm-torchknob' to the same value as your project name (which you specify in the target_link_libraries command).

At this point, attempt a build on your project to verify you have those commands set correctly. Cmake should run without errors, and you should be able to build and run the resulting project. Since you have not changed the code, merely adding libtorch to the build, the plugin will be functionally identical to the FM superknob plugin.

An important thing to note is that you can no longer use the JuceHeader.h as an include. The reason is that at the time of writing, the JuceHeader.h file calls using namespace juce, which places many things in the global namespace. Torch also puts many things in the global namespace, some of which have the same name as the ones placed there by JUCE. This causes compiler problems. So instead of including the entire JuceHeader.h you should include the individual headers from the JUCE modules folder. For example, here is how I have my includes set up in the PluginEditor.h file:

```
1  // not this:
2  //#include <JuceHeader.h>
3  //But this:
4  #include <juce_gui_basics/juce_gui_basics.h>
5  #include <juce_audio_utils/juce_audio_utils.h>
```

You will need to figure out which headers to include to access certain parts of the JUCE API. The JUCE documentation tells you where each class is defined.

5.10 Check your libtorch and JUCE project

Before moving on with the build, you should conduct a quick test to verify that you can use libtorch components in your JUCE project. Add the following code to the top of the PluginEditor.cpp file:

```
#include <torch/torch.h>
```

Now add this code to the constructor in the PluginEditor.cpp file:

```
torch::Tensor tensor = torch::rand({1, 1});
std::cout << "JUCE and torch " << tensor << std::endl;
```

Note that I am using std::cout since JUCE's DBG macro cannot handle printing tensors like this. You should see some output like this:

```
JUCE and torch  0.1046
[ CPUFloatType{1,1} ]
```

5.11 Progress check

You should now have libtorch installed on your system. You should be able to create command-line libtorch projects which access the libtorch API to create data and print it out. You should also be able to combine libtorch with JUCE using CMake. You should be able to access the libtorch library from your JUCE plugin code.

6

Python setup instructions

In this chapter, I will provide instructions on how to set up your Python environment for the work in later chapters. We will only use Python for training neural networks and some analytical work. We will use C++ to build applications. When Python code is needed, I will not explain it in such great detail as the C++ code because machine learning in Python is very well covered in other books, unlike the C++ content. I will explain the principles behind the code and provide working example code in the code repository. Only some of the examples in the book require Python code.

6.1 How to install Python

There are lots of ways to install Python. If you are an experienced Python developer, you can use whatever setup you prefer, though I recommend using virtual environments, as discussed below. If you have not installed Python on your machine, here are some instructions.

6.1.1 Python on Windows

Search for 'Python Windows releases' using your preferred search engine. You should reach a page on www.python.org with a list of Python versions with different installers for Windows. The code in this book is tested up to Python 3.10, so select the Python 3.10 link entitled something like 'Download Windows installer (64-bit)'. Run the installer. You should now be able to run the python command in your preferred Windows shell, e.g., Powershell.

Try the following in your shell to verify that Python works:

```
PS C:\Users\me> python.exe
Python 3.8.2 (tags/v3.8.2:7b3ab59, Feb 25 2020, 23:03:10) [MSC v.1916
    64 bit (AMD64)] on win32
Type "help", "copyright", "credits" or "license" for more information
    .
>>> print("Lets make some brains")
```

If you see the message printed out, you are ready to go.

6.1.2 Python on macOS

macOS comes with Python pre-installed. You should be able to run python3 on the command line – note that the command is python3, not just Python. Fire up the Terminal app from the Applications/Utilities folder:

```
1  myk@Matthews-Mini ~ % python3
2  Python 3.9.6 (default, May  7 2023, 23:32:44)
3  [Clang 14.0.3 (clang-1403.0.22.14.1)] on darwin
4  Type "help", "copyright", "credits" or "license" for more information
   .
5  >>> print("Lets make some brains")
```

If, for some reason, you do not have Python installed on your Mac, you can install it with homebrew, as recommended for the CMake install previously. You probably do not need to do this if Python is installed. You can even select particular versions:

```
1  myk@Matthews-Mini ~ % brew install python@3.10
2  myk@Matthews-Mini ~ % /opt/homebrew/bin/python3.10
3  Python 3.10.13 (main, Aug 24 2023, 22:36:46) [Clang 14.0.3 (clang
      -1403.0.22.14.1)] on darwin
4  Type "help", "copyright", "credits" or "license" for more information
   .
5  >>> print("Lets make some brains")
```

6.1.3 Python on Linux

As for macOS, your Linux system should already have Python installed. Just fire it up from a shell:

```
1  myk@yogurt52 $ python3
2  Python 3.10.12 (main, Jun 11 2023, 05:26:28) [GCC 11.4.0] on Linux
3  Type "help", "copyright", "credits" or "license" for more information
   .
4  >>> print("Lets make some brains")
```

6.2 Python packages

The Python language has built-in library functions that will allow you to carry out various tasks. But if you want to do more specialised work, such as machine learning and audio processing, you must install additional packages to your machine. Python packages are equivalent to C++ libraries. I will show you how to install Python packages after I explain how to set up a virtual environment in the next section.

6.3 Virtual environments

Python virtual environments allow you to set up Python environments on your machine with particular package versions installed. The idea is that the virtual environment can be a frozen environment that is more likely to work in the future, even if you upgrade your system-wide Python install. It is common for machine learning researchers to provide package lists along with their code repositories, which allow you to set up a virtual environment that is as close as possible to the one in which they developed their code.

I recommend you do all your machine learning work in Python virtual environments. At least when you are working on your own machine. Here are the commands to set up and activate a Python virtual environment.

Windows:

```
1  python.exe -m venv C:\Users\myuser\myvenv
2  cd myvenv
3  .\Scripts\activate
```

macOS and Linux:

```
1  python3 -m venv /path/to/my/mypythonenv
2  source /path/to/my/mypythonenv/bin/activate
```

6.4 Anaconda's virtual environments

Some people prefer to use Anaconda to control their Python environments. In that case, you need to run different commands to create and activate the virtual environment. Alternatively, you can use the Anaconda UI to configure things. I will provide the Anaconda commands here in case they are helpful, but I have experimented with Anaconda, and I prefer my lower-level, straightforward Python setup.

```
1  conda create -n yourenvname
2  conda activate yourenvname
```

6.5 Installing packages in your venv

You can start installing packages once you have created and activated a virtual environment. You can install packages using the pip command. Here is a decent list of starter packages I recommend (assuming you are in your virtual environment):

```
1  pip install scipy numpy matplotlib jupyter ipython librosa pandas
```

Here is a simple Python script that will use some of these packages to generate a 400Hz sine tone and save it to disk as a WAV file:

```
1  import numpy as np
2  from scipy.io.wavfile import write
3  freq = 400
4  x = np.sin(np.arange(0, np.pi * freq*2, (np.pi * freq * 2) /44100))
5  write('sine_400Hz.wav', 44100, x * 0.5)
```

6.5.1 Problems with packages

It is common practice for AI researchers and AI music researchers to provide Github repositories, along with their papers, with the code they used to carry out the experiments. They will even provide the dataset and a pre-trained model if you are lucky. My experience is that even when researchers provide a list of packages with version numbers, I still encounter problems setting up environments to run their code. But the problems generally relate to the order in which packages are installed. Installing a package also attempts to install its dependencies, which might result in a different version of a dependency than the one the researchers used being installed. Sometimes, you must manually install the packages in non-alphabetical order to get the correct versions, which is not fun. As you move beyond this book and experiment with machine learning software repositories you find for yourself, you may encounter this problem, but at least I have given you some hints on how to solve it.

6.6 IPython and Jupyter notebooks

You may have spotted in my list of packages that I included IPython and Jupyter. These packages provide enhanced Python environments. Firstly, run Python on the command line, and you will enter a more powerful interactive Python shell:

```
1  python
2  Python 3.10.12 (main, Jun 11 2023, 05:26:28) [GCC 11.4.0]
```

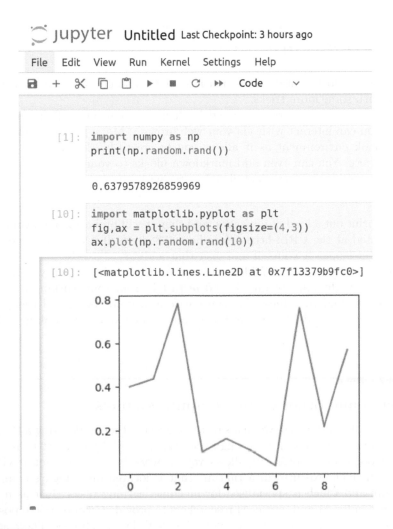

FIGURE 6.1
A Jupyter notebook in action. There are cells containing Python code which you can execute. If you trigger a plot command, the plot will be embedded in the worksheet.

```
3 Type 'copyright', 'credits' or 'license' for more information
4 IPython 8.16.1 -- An enhanced Interactive Python. Type '?' for help.
5 In [1]:
```

Try running an import command and then querying its documentation:

```
1 from scipy.io.wavfile import write
2 ?write
```

You can access documentation within the shell. Go and find yourself an Ipython tutorial to learn some more tricks.

Jupyter notebooks is a browser-based Python environment. It fires up a web application you can interact with via your web browser. Many people like to work in the notebook environment as it allows you to create blocks of code, display graphs inline, etc. You can even add markdown blocks to your notebook. To run the system, assuming you have installed Jupyter with pip:

```
1 jupyter-notebook
```

This will print out a load of messages, and then it will probably open your web browser, pointed at the URL: http://localhost:8888/tree. If it does not open the web browser, just open it yourself and visit that address.

You will see a listing of the folder you were in in your shell when you ran the command. You might want to quit it and restart it from your working folder to make it easier to find your files. You can create a new notebook and then start entering and running code. See figure 6.1 for an example of a notebook in action.

6.7 Colab and other cloud compute systems

Jupyter notebooks in the form we just saw them execute their code using a Python runtime on your local computer (though they can potentially connect to remove runtimes). Google's colaboratory[1] allows you to work in the notebook environment, but your code executes in a runtime that is located on Google's compute infrastructure and which is specialised for machine learning tasks. Just as it has become common for AI and music-AI researchers to provide Github code repositories, there is a more recent trend to provide Google Colab notebooks. These allow other researchers to immediately execute the code without any installation and using a GPU-accelerated back-end. You can upload your locally developed Jupyter notebook into colab and execute it. You can also connect to your Google Drive storage from the notebook to store training data and model training checkpoints. More on training models later.

---

[1]https://colab.research.google.com/

6.8   Progress check

At this point, you should have Python installed on your computer. You should understand what a virtual environment is and know how to create one. In your virtual environment, you should be able to install Python packages and then import and use them in a Python script. You should also have installed and experimented with Jupyter notebooks, which are excellent for quick experimentation and visualisation. You should be aware of Google Colabs and understand that it can be used to execute Jupyter notebooks or regular Python code in the cloud with GPU-accelerated machine learning capabilities.

7

Common development environment setup problems

This section provides solutions for common problems people encounter when setting up their development environment and building applications.

Problem The IDE does not recognise anything about JUCE when I generate a project from Projucer.

Solution This can happen if you generate a project with Projucer and later move the JUCE files to a different location. To verify your setup, try generating a new project from Projucer, checking the module setup and then building it. If that works, you can return to your non-working project in Projucer and determine what differs from the newly generated project.

Problem Visual Studio complains about the v143 (or higher) toolkit not being installed when you try to build a project.

Solution Re-run the Visual Studio Installer and select the C++ packages with 143 (or higher) in their titles. Or you can choose a different platform toolset from Projucer's Visual Studio exporter configuration.

Problem The IDE build completely fails on a newly created Projucer project.

Solution This can happen if you do not have the appropriate components installed with your IDE. For Visual Studio, you need the components shown in the instructions earlier in this section. For Xcode, you need to install the Xcode command line tools to work with C++. For Linux, you need the build-essentials (Debian) or equivalent for your distribution.

Problem Building with libtorch / JUCE combination complains that a reference to 'nullopt' is ambiguous.

Solution This is caused by including the complete JuceHeader. JuceHeader calls 'using namespace juce', which puts all JUCE definitions and such in the 'global' namespace. libtorch also puts some things there that then clash with the JUCE ones. The solution is to include the individual JUCE headers you need instead of the complete JuceHeader.h file.

Problem The compiler does not seem to understand basic code syntax, such as initialiser lists on constructors.

Solution I have observed this when building projects on a Mac using Visual Studio Code. It can be caused by the default C++ version being pre- C++11.

Before C++11, initialiser lists and other syntax were not part of the standard. The solution is to tell the build tools which C++ version you want to use. Add the following to your CMakeLists.txt file:

```
set_property(TARGET project-name PROPERTY CXX_STANDARD 14)
```

You should change project-name to the project name you set in the CMake project command.

8

Basic plugin development

In this chapter, you will build a simple sine wave synthesizer plugin. This process will show you some essential aspects of plugin development. You will learn about the processBlock function and how to fill up a buffer with numbers representing the signal you want your plugin to generate. You will learn about user interface widgets and how to edit your plugin's user interface. This will involve implementing event listeners and layout code. At the end of the chapter, you will have a simple plugin up and running.

8.1 Constant tone sinewave synth

Now that you have a working system to create JUCE plugin projects using CMake, you are ready to implement some basic plugins. The first plugin you will work on is described in section 39.2.3 on page 329 in the repository guide, and it is a sine wave synthesizer.

The first version of the plugin will generate a constant sine tone with a constant frequency. Start with the CMake/ JUCE plugin project seen in the previous section. The repository guide describes this project in section 39.2.2. Adjust the project name, plugin name and so on as suggested in section 4.10.2. Now you are ready to begin.

To get the sine tone plugin up and running you will need three variables and a function. Add three variables of type double to the private section of the class defined in PluginProcessor.h:

```
...
private:
    double phase;
    double dphase;
    double frequency;
...
```

Now add a function prototype (the specification for a function) to the private section of the class defined in PluginProcessor.h:

```
...
private:
    double getDPhase(double freq, double sampleRate);
...
```

Open up PluginProcessor.cpp and find the constructor. You will initialise the phase, dphase and frequency variables here:

```
TestPluginAudioProcessor::TestPluginAudioProcessor
...
{
    phase = 0;
    dphase = 0;
    frequency = 440;
}
...
```

Or, if you want to use a more modern C++ style, you can use an initialiser list:

```
TestPluginAudioProcessor::TestPluginAudioProcessor
...
) // at the end of the brackets
 // put the initialiser list
 : phase{0}, dphase{0}, frequency{440}
{
    // then remove the phase = 0 etc.
    // things from inside the constructor
}
...
```

Then the implementation of the getDPhase function tells us how much the phase of the sine generator changes per sample and allows us to synthesize a sine tone with the correct frequency for the current audio system configuration.

```
1  double TestPluginAudioProcessor::getDPhase(double freq, double
       sampleRate)
2  {
3      double two_pi = 3.1415927 * 2;
4      return (two_pi / sampleRate) * freq;
5  }
```

Let's clarify what that code is doing. The sine wave gets through its complete cycle in 2π. So if you compute the output of the sine function with values rising from 0 to 2π (about 6.3) and plot it, you will see the complete sine wave. The input to the sine function is called the phase. The top half of figure 8.1 illustrates how the sine function varies with phase.

But digital audio systems are discrete – they go in steps. In fact, we have 'sampleRate' steps in one second, e.g. 44,100 steps. So to complete the sine wave in one second, we need 44,100 values. So we slice 2π into 44,100 steps, which is how much the phase changes in each step. For a 1Hz sine wave with the audio system

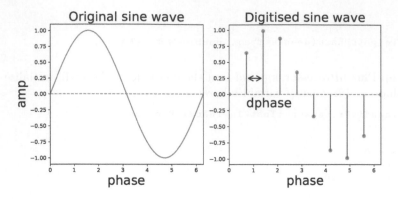

FIGURE 8.1

dphase depends on the sample rate (the space between the samples) and the frequency (how fast you need to get through the sine wave).

sample rate set to 44,100Hz, dphase is $\frac{2\pi}{44100}$. That is the first part of line 4 in the getDPhase function. The lower half of figure 8.1 illustrates a discrete version of the sine function.

What happens if the frequency is higher? Higher frequency means the sine wave goes through its cycle faster so dphase is higher. Double the frequency, double the dphase. So that is why we scale the 1Hz dphase by the frequency in the second part of line 4 in the getDPhase function.

8.2 Lifecycle of a plugin

When you instantiate a plugin in a plugin host, the host calls the plugin's functions in a particular order. During this process, the plugin gains access to the audio system provided by the plugin host. It works like this:

1. PluginProcessor's constructor is called. At this point the audio system is not available so you cannot query the sample rate or block size.

2. PluginProcessor's prepareToPlay function is called. The audio system is ready; you now know the sample rate and block size.

3. PluginProcessor's processBlock function is called repeatedly. This is where you generate or process audio and put the results into the buffer sent to processBlock.

4. PluginProcessor's releaseResources function is called. At this point the audio system is shutting down.

5. PluginProcessor's destructor is called.

The point at which you can compute the dphase value is in prepareToPlay because that is when you have information about the audio system configuration (the sample rate).

Add this to the prepareToPlay function in PluginProcessor.cpp:

```
dphase = getDPhase(frequency, sampleRate);
```

8.3   The processBlock function

Processing audio in blocks is a common paradigm for audio programming. You can think of it like a single carriage train entering a station. The train stops, and passengers are processed. Some get on, some get off. Later, another train arrives. Your plugin is the station, and the passengers are the audio buffer. Depending on the type of plugin you are creating, you will need to do different things with the audio buffer or block. If you are making an audio effect that processes an existing signal, you will take the data from the buffer, change the numbers, and then write the adjusted numbers back to the block. If you are building a synthesizer, you will ignore the numbers arriving in the block and just synthesize your own numbers. A plugin synthesizer is illustrated in figure 8.2. It is essential to process the block quickly, as ultimately, the audio block will travel to the audio output. The audio output requires a constant stream of blocks to create a continuous stream of audio. There will be audio dropouts if you do not process your block quickly enough.

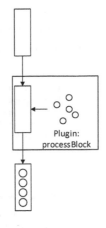

FIGURE 8.2
A synthesizer plugin loads data into the incoming blocks.

The processBlock function has the following signature:

```
void processBlock (juce::AudioBuffer<float>&,
                   juce::MidiBuffer&)
```

You can see that it receives an audio buffer and a MIDI buffer. In the case of a synthesizer, the audio buffer provides a place to write the audio output of the synthesizer. The MIDI buffer provides a place to receive MIDI messages that the synthesizer needs to process.

For the sine synthesizer, the processBlock function needs to write a sine tone into the buffer. Remember that this function will automatically be called repeatedly by the audio host so that it can receive audio from your plugin. For now, we will use the std::sin function to generate the sine wave. There are more efficient ways to do it than this one, but it will be efficient enough for now. We will start with a one-channel sine tone – noting that the plugin is probably running in stereo.

Usually, the default processBlock code that Projucer generates for a blank project looks something like this:

```
1 juce::ScopedNoDenormals noDenormals;
2 auto totalNumInputChannels  = getTotalNumInputChannels();
3 auto totalNumOutputChannels = getTotalNumOutputChannels();
4 for (auto i = totalNumInputChannels; i < totalNumOutputChannels; ++i)
    {
5     buffer.clear (i, 0, buffer.getNumSamples());
6 }
```

Line 1 looks a little odd but essentially it prevents extremely small floating point values (denormals) from causing problems. Then the code iterates over the channels in the sent buffer and clears them out so they are silent. Add the following code after those lines in the default processBlock code:

```
1 for (int channel = 0; channel < totalNumOutputChannels; ++channel)
2 {
3     if (channel == 0){
4         auto* channelData = buffer.getWritePointer (channel);
5         int numSamples = buffer.getNumSamples();
6         for (int sInd=0;sInd < numSamples; ++sInd){
7             channelData[sInd] = (float) (std::sin(phase) * 0.25);
8             phase += dphase;
9         }
10     }
11 }
```

Read that code carefully – what do you think it is doing? Can you make the sine tone louder? Be careful if you are wearing headphones. Try running your plugin in the JUCE AudioPluginHost.

Challenge: can you make the sine tone work in stereo? You could use an array of phases, one for each channel, or can you do it with a single phase variable?

8.4   Changing frequency with a slider

If you have worked in music technology for some time, you will undoubtedly have asked yourself, "Is this thing on?". Now you have a plugin that is always on. If that sounds like a valuable thing to have lying about, store a copy of your plugin code with the constantly playing sine tone, and prepare a new copy where you can continue working on it.

You will now make the sine tone interactive by adding a GUI control for the frequency. There are two parts to this: adding a widget to the user interface and then adding the code to respond to the widget. Adding a widget is a three-stage process:

1. Add a variable to the private area of the PluginEditor class in PluginEditor.h:

```
private:
...
juce::Slider freqControl;
```

2. Call addAndMakeVisible in the PluginEditor constructor in PluginEditor.cpp:

```
// our UI class inherits this function
// from the Component class
addAndMakeVisible(freqControl);
```

3. Call setBounds on the slider in the PluginEditor resized function in PluginEditor.cpp:

```
// set the width to the width
// of the whole UI and
// the height to half
// the height of the whole UI
freqControl.setBounds(0, 0,
                 getWidth(), getHeight()/5);
```

Carry out these steps, then build and run it. You should see the slider displayed on the user interface. The JUCE Slider class reference can be found here[1]. Now you have the slider displaying, you need to respond to it. You must implement the SliderListener interface in your PluginEditor to respond to a slider. This is a four-step process:

1. Specify that your class inherits from the Slider::Listener. In PluginEditor.h:

```
class TestPluginAudioProcessorEditor   :
public juce::AudioProcessorEditor,
public Slider::Listener
```

2. Provide prototypes for the pure virtual functions[2], In PluginEditor.h:

```
void sliderValueChanged (Slider *slider) override;
```

---

[1]https://docs.juce.com/master/classSlider.html
[2]https://docs.juce.com/master/classSlider_1_1Listener.html

3. Provide implementations for the pure virtual functions on the SliderListener, in PluginEditor.cpp:

```
1  void TestPluginAudioProcessorEditor::sliderValueChanged (
       Slider *slider)
2  {
3      if (slider == &freqControl){
4          // get the slider value and do something
5          DBG("Slider value " << slider->getValue());
6      }
7  }
```

4. Now tell the slider that we want to listen to it: in the PluginEditor.cpp, constructor:

```
freqControl.addListener(this);
```

If you run this code, you should see messages printed to the console when you move the slider. The next step is to change the parameters on the synthesis system so it changes the frequency of the sine tone according to the value set on the slider. You must establish communication between the PluginEditor and the PluginProcessor to achieve that. First, you need a public function on the PluginProcessor, which the PluginEditor can call to change the frequency. Add the following function signature to the public section of PluginProcessor.h:

```
void updateFrequency(double newFreq);
```

Can you remember the name of the variable that we need to change to make the sine wave go faster? Can you remember how we set the initial frequency for the sine wave? Check out the code in prepareToPlay. You need to do the same thing in update frequency. So, in PluginProcessor.cpp, add an implementation for updateFrequency:

```
1  void TestPluginAudioProcessor::updateFrequency(double newFreq)
2  {
3      frequency = newFreq;
4      dphase = getDPhase(frequency, getSampleRate());
5  }
```

We stored the updated frequency in case we needed it later and changed the dphase variable to reflect the new frequency. The final step is to call this function from the PluginEditor when the slider moves, so back to PluginEditor.cpp:

```
1  void TestPluginAudioProcessorEditor::sliderValueChanged (Slider *
       slider)
2  {
3      if (slider == &freqControl){
4          // get the slider value and do something
5          DBG("Slider value " << slider->getValue());
6          audioProcessor.updateFrequency(slider->getValue());
7      }
8  }
```

Notice the addition of line 6, where we call updateFrequency on the audioProcessor variable. You can see that the PluginEditor has access to public functions on the PluginProcessor via this audioProcessor variable. Once you have this code in place, you should hear the frequency of the sine wave change when you move the slider. In case you have problems, section 39.2.4 describes a working solution.

8.5 Changing frequency from MIDI input

The next thing you should do with the sine plugin is to change the frequency in response to MIDI input. This provides an opportunity to examine how plugins receive and process MIDI data. To achieve this, we need to check some plugin details and update the processBlock function. First of all, check the implementation of the acceptsMidi function.

```
bool acceptsMidi() const override;
```

You should see that it looks like this:

```
#if JucePlugin_WantsMidiInput
return true;
#else
return false;
#endif
```

So, instead of simply returning true or false, it checks a property set in the CMakeLists.txt file. Look in the CMakeLists.txt file for this line:

```
NEEDS_MIDI_INPUT TRUE
```

You can simply return true or false, but the polite JUCE way is to edit CMakeLists.txt. If you update CMakeLists.txt, you might need to regenerate your IDE project. The IDE should auto-detect the change and ask if you want to reload. Now your plugin is ready to receive MIDI. Let's take a look at the processBlock function signature:

```
void processBlock (juce::AudioBuffer<float>&, juce::MidiBuffer&)
    override;
```

You can see it receives a MidiBuffer object. This will contain any incoming MIDI data your plugin might want to process. We shall start by just printing out the messages. Add this code which uses a foreach loop to iterate over the incoming MIDI messages to the end of your processBlock function:

```
for (const auto metadata : midiMessages){
    auto message = metadata.getMessage();
    DBG("processBlock:: Got message " << message.getDescription());
}
```

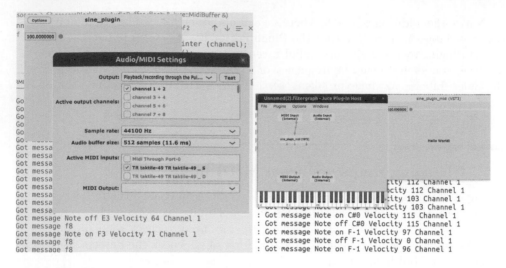

FIGURE 8.3
Printing descriptions of MIDI messages coming into a plugin in Standalone mode from a USB controller keyboard (left) and MIDI coming from an on-screen piano keyboard in AudioPluginHost (right).

Now you need to send it some MIDI. If you have a MIDI input device such as a USB controller keyboard, you can run the plugin in Standalone mode, select the USB device from the options and check if messages are printed out when you press the keys. Figure 8.3 illustrates this scenario. You can also use the AudioPluginHost to send MIDI to your plugin if you do not have a USB MIDI controller. Create the graph shown in figure 8.3, where the MIDI input block in the host is wired to the MIDI input of the plugin. You might need to scan for new plugins to add your plugin to the list. Click on the buttons on the on-screen piano keyboard.

Now we have MIDI coming into the plugin, we need to decide how to respond to it. Going back to our MIDI parsing code, let's put some logic in there to update the frequency in response to a note-on message:

```
1  for (const auto metadata : midiMessages){
2      auto message = metadata.getMessage();
3      DBG("processBlock:: Got message " << message.getDescription());
4      if (message.isNoteOn()){
5          // call updateFrequency with the midi note numbers
6          // converted to hz
7          updateFrequency(
8              juce::MidiMessage::getMidiNoteInHertz(
9                  message.getNoteNumber())
10         );
11         break;
12     }
13 }
```

Read the code carefully. You will see that it uses the isNoteOn function to check for MIDI note-on messages. It uses the getMidiNoteInHertz function to convert the MIDI note from the range 0–127 to a frequency value. On line 8, it updates the dphase value using the frequency extracted from the note-on message.

8.6 Challenge: amplitude control

The next step is to switch the sine tone on and off in response to MIDI messages. The following logic will work: If you receive a note-on message, set the frequency to the correct value for the note and set amplitude to max. If you receive a note-off message: set the amplitude to zero. Can you add some code to implement this logic? Section 39.2.5 in the repository guide demonstrates the solution to this. Can you implement a proper envelope that fades in and out? And some UI controls for it?

8.7 Progress check

You should now be able to create a simple plugin and implement basic sound synthesis in its audio process class. You should know how to add graphical user interface elements to your plugin and implement the appropriate event listeners so the plugin can respond to user input via its user interface. You should be able to process incoming MIDI events and use note messages to set the frequency of the synthesis algorithm. There were some challenges set, which you will undoubtedly learn from if you attempt them; for example, implementing an amplitude envelope to fade the sound in and out.

9

FM synthesizer plugin

In this chapter, you will develop a plugin with a more advanced sound synthesis algorithm: frequency modulation synthesis (FM). This FM synthesis plugin will be helpful for later work in the book, where you will use machine-learning techniques to control the plugin. I will show you how to implement a two-oscillator FM synthesis algorithm with a modulator and a carrier oscillator. I will explain how the modulation index and depth variables change the timbre. You will add a slider-based user interface to the plugin, allowing you to control the synthesis engine. I will then show you how to expose plugin parameters to the host environment, which is one way that plugins integrate more deeply with plugin hosts. Then, we will see how to use those plugin parameters in the synthesis algorithm and how to control them via the user interface. At the end of the chapter, you will have a useful monophonic FM synthesis plugin.

9.1   FM synthesis algorithm

In the previous chapter, you saw how to create a simple synthesiser plugin that generated a sine tone. Sine tones are nice, clean sounds; if you add many of them together, you can create rich timbres. The problem with this kind of additive synthesis, where you add sine tones together, is that it is computationally expensive and difficult to control. Frequency modulation (FM) synthesis aims to solve both problems by generating complex timbres computationally efficiently with few control parameters. For music technology history fans, there is an interesting story to look into relating to FM synthesis and how it moved from John Chowning's research work into commercial products made, most famously by Yamaha but also by the New England Digital Corporation in their Synclavier series.

The basic FM algorithm works by using one oscillator to control the frequency of another at audio rate. You are probably familiar with the idea of an LFO modulating pitch, where the pitch can be heard to go up and down like a police car siren. FM is like that, except the frequency moves up and down much faster. The result of this fast frequency modulation is that the tone of the modulated oscillator (the carrier) becomes richer, with additional frequencies appearing in its spectrum. The following expressions specify the parameters and components of a simple FM synthesis algorithm which involves two oscillators, with example values for the parameters:

$$
\begin{aligned}
f_0 &= 400 & &\text{base frequency} \\
m_i &= 0.5 & &\text{modulation index} \\
m_d &= 100 & &\text{modulation depth} \\
m_f &= f_0 m_i & &\text{the modulator frequency} \\
m &= m_d \sin(2\pi m_f t) & &\text{the modulator signal at time t} \\
c_f &= f_0 + m & &\text{the carrier frequency} \\
c &= \sin(2\pi c_f t) & &\text{the carrier signal at time t}
\end{aligned}
$$

$$(9.1)$$

9.2   FM synthesis in code

Now you have seen the algorithm, let's convert it into code in a plugin. You can start from example 39.2.6 in the repo guide, a sine oscillator plugin with an envelope generator that responds to MIDI notes. Edit CMakeLists.txt with a new PRODUCT_NAME and PLUGIN_CODE value. Run a test build to verify things are working.

The plugin has variables to manage the amplitude of its output plus the phase and frequency of its single oscillator, which we can refer to as the carrier oscillator. We need to add variables to manage the phase and frequency of the modulator oscillator. So open up PluginProcessor.h and add the following to its private section:

```
double mod_phase;
double mod_dphase;
double mod_index;
double mod_depth;
```

Then initialise them in the initialiser list, which you can find just above the constructor body in PluginProcessor.cpp – I am setting the index and depth as I did in the expressions above:

```
    ...   mod_phase{0}, mod_dphase{0}, mod_index{0.5}, mod_depth{100}
// constructor body follows
{}
```

Now to set mod_dphase based on the frequency (f_0) and mod_index (m_i) variables. In PluginProcessor.cpp's updateFrequency function, you will see lines that set up the original carrier oscillator, so add some more for the modulator oscillator:

```
1 frequency = newFreq;
2 dphase = getDPhase(frequency, getSampleRate());
3 // the mod_index is a multiplier on the base frequency:
4 mod_dphase = getDPhase(frequency * mod_index, getSampleRate());
```

We are about to achieve rich FM tones. The next step is to change the synthesis algorithm in PluginProcessor.cpp's processBlock function. You should see something like this, which writes the output of a sine function to the channel data buffer:

```
1 channelData[sInd] = (float) (std::sin(phase) * amp);
2 phase += dphase;
```

Change that block to this:

```
1 channelData[sInd] = (float) (std::sin(phase) * amp);
2 mod = std::sin(mod_phase);
3 mod *= mod_depth;
4 dphase = getDPhase(frequency + mod, getSampleRate());
5 phase += dphase;
6 mod_phase += mod_dphase;
```

Now build and run – you should hear a richer, bassy tone. This is because the modulator index is 0.5, so the modulation ends at half the base frequency. You can experiment with different values for mod index and mod depth now, or you can move on to the next section, wherein we add slider controls for those parameters.

9.3 Slider controls

You can easily add sliders to control the mod index and mod depth. Use the following procedure, which you saw the full code for in the previous chapter:

1. Add two Slider fields to the private member variables in PluginEditor.h
2. Call addAndMakeVisible(slider) with the slider in the constructor in PluginEditor.cpp
3. Call setRange on the sliders in the constructor of PluginEditor.cpp – for mod index, use the parameters 0.25, 10, 0.25, and for mod depth, use the parameters 0, 1000.
4. Call slider addListener(this) in the constructor in PluginEditor.cpp
5. Call setBounds on the slider in the resize function in PluginEditor.cpp

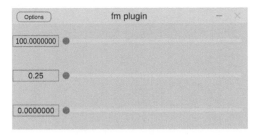

FIGURE 9.1
Simple FM plugin with sliders for frequency, modulation index and modulation depth.

6. Update the sliderValueChanged function in PluginEditor.cpp to detect when your new sliders are changed – add two new if blocks in there.

Now, you need to allow the user interface component of the plugin to tell the audio processor part to update mod_depth and mod_index. You can do this by adding two 'setters' to the public section of PluginProcessor.h:

```
void setModIndex(double newIndex);
void setModDepth(double newDepth);
```

Then implement in PluginProcessor.cpp:

```
void TestPluginAudioProcessor::setModIndex(double newIndex)
{
    mod_index = newIndex;
    updateFrequency(frequency);
}
void TestPluginAudioProcessor::setModDepth(double newDepth)
{
    mod_depth = newDepth;
}
```

Note that I called updateFrequency in setModIndex because setting the mod index requires that the mod_dphase is updated to reflect the new index. Remember that the mod_index dictates the frequency of the modulator. The final set is to hook up the sliders so they call these new functions on the audio processor. In PluginEditor's sliderValueChanged function:

```
if (slider == &modIndexControl){
    audioProcessor.setModIndex(slider->getValue());
}
if (slider == &modDepthControl){
    audioProcessor.setModDepth(slider->getValue());
}
```

If you managed to get this all working, you should see a user interface like

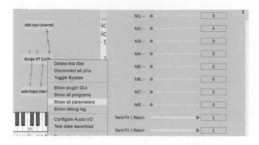

FIGURE 9.2
Showing plugin parameters for the Surge XT synthesiser using AudioPluginHost.

figure 9.1. The repo guide provides the code for this working FM plugin in example 39.2.7.

9.4 Plugin parameters

One part of the underlying plugin APIs we have yet to examine is the plugin parameters. The concept of plugin parameters is that any plugin can expose a list of parameters to its host. With those parameters, the host can carry out tasks like parameter automation. Parameters are also the basis for presets. Think about the FM plugin – it has two parameters (if we ignore the frequency control). The modulation index and the modulation depth. Currently, the only way to change those parameters is via the plugin's user interface. A host cannot see the parameters for a plugin.

You can see the parameters your installed plugins expose to the host using the AudioPluginHost. Figure 9.2 shows how you can right-click on a plugin in AudioPluginHost and choose to view its parameters as a simple slider user interface. The parameter user interface is not ideal for editing hundreds of settings, but it shows what the host can see about the plugin's parameters. If you do the same thing to your FM plugin, you will see it does not expose any parameters. Let's fix that!

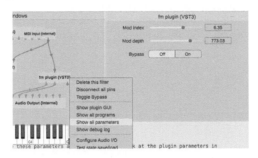

FIGURE 9.3
Showing plugin parameters for the FM plugin using AudioPluginHost.

9.5 Adding parameters to the FM plugin

Add the following two fields to the private section in PluginProcessor.h:

```
1 juce::AudioParameterFloat* modIndexParam;
2 juce::AudioParameterFloat* modDepthParam;
```

Now in the constructor for the audio processor in PluginProcessor.cpp, set up the parameters:

```
1 modIndexParam =
2     new juce::AudioParameterFloat("ModInd", // Parameter ID
3                                   "Mod index", // Parameter name
4                                   0.0f,   // Minimum value
5                                   10.0f,  // Maximum value
6                                   0.5f);  // Default value
7
8 addParameter(modIndexParam);
9 modDepthParam =
10     new juce::AudioParameterFloat("ModDep",
11                                   "Mod depth",
12                                   0.0f,
13                                   1000.0f,
14                                   100.0f);
15 addParameter(modDepthParam);
```

You should see these parameters appearing when you look at the plugin parameters in AudioPluginHost, as shown in figure 9.3. But how do you make those parameters influence the synthesis algorithm? Add the following to processBlock in PluginProcessor.cpp before the loop that fills up the channel buffer:

```
1 mod_dphase = getDPhase(frequency * static_cast<double>(*modIndexParam
     ), getSampleRate());
2 mod_depth = static_cast<double>(*modDepthParam);
```

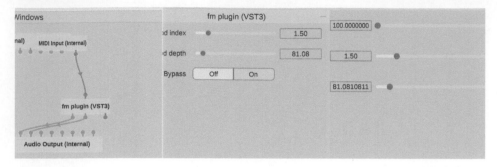

FIGURE 9.4
Showing the custom UI for the FM plugin (right), the auto-generated parameter UI (middle) and AudioPluginHost (left).

Remember the mod_dphase is the speed the modulator oscillator goes at (so it dictates its frequency). We are reading in the modIndexParam, using some fancy syntax to convert it to a double, and then using it to update the mod_dphase. Following that, we set up the mod_depth variable using the modDepthParam parameter. This is done as mod_depth can be used directly from the modDepthParam parameter instead of being calculated from it as for mod_dphase.

Note that the original GUI sliders we set up do not influence the algorithm any more. You can change the two setters we created earlier in PluginProcessor.cpp (setModIndex and setModDepth), so they change the parameters instead of the variables.

```
void TestPluginAudioProcessor::setModIndex(double newIndex)
{
    *modIndexParam = (float) newIndex;
}
void TestPluginAudioProcessor::setModDepth(double newDepth)
{
    *modDepthParam = (float) newDepth;
}
```

Figure 9.4 shows the result, which is that the plugin has two possible interfaces: 1) a custom user interface, defined in PluginEditor.h, which has the original set of controls we created and 2) a generic interface that the host auto-generates by querying the parameters on the plugin. You can find the completed FM plugin in example 39.2.8 in the repo guide. I will make one comment here about efficiency. Calling std sin might not be the most efficient way to generate a sine signal – JUCE has classes in its library that provide better oscillators. But for now, with two oscillators running like that, it is efficient enough and makes the code more straightforward.

9.6 Progress check

Excellent stuff – you should now be getting to grips with plugin development. If you are wondering why we are making all these plugins but not doing any AI or machine learning, don't worry; I want you to gain some familiarity with developing plugins and their inner workings before we pile on additional layers of complexity. There is plenty of that content coming up later. Also, the FM plugin will come in handy later in the book when we work on the meta-controller.

9.7 The end of the first part of the book

Congratulations – you have reached the end of the book's first part. You have worked your way through some quite complex setup procedures, and I expect you have encountered and fought your way through strange error messages, segmentation faults and syntax errors. Still, if you can now build plugins and run them in your host, you have a working Python development environment, and you can integrate libtorch with your C++ programs, you are ready to crack on with the rest of the book.

Part II
ML-powered plugin control: the meta-controller

10

Using regression for synthesizer control

This chapter introduces regression, a statistical technique to estimate values between known data points. The chapter starts with 2D linear regression, fitting a line to known data points. Then, it expands the concept to higher dimensions to enable one-to-many mappings. I explain how this relates to sound synthesiser control. I will show you how neural networks can implement the same line equation as linear regression, where weights and biases are the parameters for the network, and learning is the process of estimating suitable values. At the end of the chapter you should have an understanding of how linear regression and simple neural networks work.

10.1   What does the meta-controller do?

FIGURE 10.1
The meta-controller uses regression to control other plug-ins.

Now that you have completed installing, testing and familiarising yourself with the book's core tools, you will work towards the first serious example project: the meta-controller. The meta-controller is an innovative tool for synthesizer programming. It allows you to control multiple parameters of a synthesizer using one or more controls, for example a single slider can control tens or hundreds of parameters. This means synthesizer users can move through the sound space of any plugin synthesizer more dynamically than they can via conventional synthesizer programming techniques. The meta-controller uses interactive machine learning to allow users to create complex linear or non-linear mappings between arbitrarily dimensional controls and synthesis parameters.

But before you dive in with the meta-controller project, I would like to explain the background machine learning technique it uses. The technique is called regression, and we'll do it using a neural

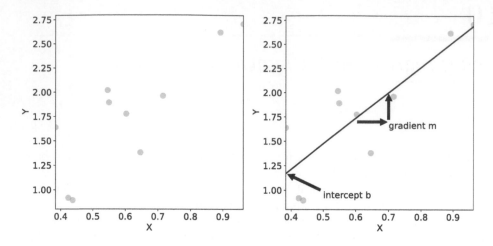

FIGURE 10.2
Linear regression finds the straight line that best fits some data. Important features of the line are the point at which it intercepts the y-axis and the slope gradient.

network. The aim for this chapter therefore is to explain what regression is and how to do it using neural networks.

10.2   Linear regression and synthesizer control

The meta-controller is an AI-enhanced synthesizer programmer which allows you to use simple controls to move through the sound space of any plugin synthesizer more dynamically than you can via conventional synthesizer programming techniques.

The meta-controller uses a technique called regression. Regression is a term from statistics that most people encounter when they learn about making predictions from graphs. The idea is that you have data for several points on a graph, but you want to know what happens between or beyond those points, so you can estimate the value at a given position for which you do not have an actual data point. Linear regression is perhaps the most straightforward method to estimate what happens between and beyond the known data points. Linear regression involves fitting a line to the available points in the best way possible.

The logical next question is – how do you work out where to put the line? First, you can consider that you can specify a line as follows:

$y = mx + b$

y is the y coordinate, x is the x coordinate, m is the gradient of the slope and c is the intercept on the y axis. These values are illustrated in Figure 10.2. So the main job of a linear regression function is to compute m and b. If you have m and b, you can calculate y for any x you like. The best line will have the smallest total error between itself and the set of known data points. If the points are not on a straight line, the best line will not go through all the points, so the linear regression function has to minimise the error between the points and the line. Once you have estimated the intercept and gradient values a and m that define the line, you can estimate any y given an x value. Simple!

10.2.1 Regression or interpolation?

Regression and interpolation are related processes that aim to determine what the data does between the known points. The difference is that interpolation requires that the resulting line goes through all the data points, and regression does not. Simple linear interpolation is not suitable for noisy data as a single line cannot pass through all the data points. There are more complex forms of interpolation that can handle more complex data, but for now, we will focus on the more 'heuristic' regression.

10.2.2 Regression and synthesizer control

What does this have to do with synthesizer parameter control, though? As mentioned above, the meta-controller lets you control several parameters from a single controller. If the meta-controller worked like the simple linear regression example above, it would estimate one parameter (the y value) given the control value (the x value). That is probably pointless – it is not much use to map one parameter to one other parameter. What we want to do with the meta-controller is map one parameter (e.g. a single slider control) to control many parameters on the synthesizer. To do that, we need to go into higher dimensions. That sounds mind-bending, but all it means is that we are estimating more than one line. So the slider value (x) maps to multiple lines or multiple parameters. Each line has its own intercept point and gradient, allowing differently calibrated control of each parameter. Figure 10.3 illustrates how, with two lines, we can estimate values for two parameters given a single input. This mapping can be referred to as one-to-many, as one value controls several others.

One-to-many mapping is already possible in commercial synthesizers. The 'Super-knob' on the Yamaha ModX synthesizer can be mapped to multiple parameters. You can map a single knob control (the Super-knob) such that it controls

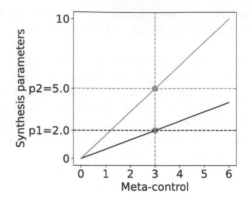

FIGURE 10.3

Linear regression with two lines, allowing the estimation of two parameters given a single 'meta-controller' input control. The x-axis represents the control input and the y-axis shows the settings for the two parameters the control is mapped to.

several other parameters simultaneously. Other synthesizers permit this kind of mapping with varying degrees of sophistication; for example, my Access Virus C synthesizer allows me to map two knobs on the control panel to various synthesis parameters via its mod matrix feature. But the Access Virus only allows me to specify the multiplier m and not the intercept point (i.e., the control range over the other parameters). Have you used a synthesizer which provides this kind of meta-control? If you want to try it out, go back to the FM synthesis example plugin and add a new controller that changes the modulation index and modulation at the same time.

The meta-controller is a relative of the Super-knob, but it is more powerful and intuitive. More powerful because the meta-controller allows a user to create linear or non-linear mappings, unlike the more common linear mappings. And it can map between arbitrarily dimensional controls, meaning you can map an x,y pad to control 10 parameters, or a single slider to control 100 parameters, or any other combination. This is many-to-many mapping. The meta-controller is more intuitive because it uses a 'learn by example' workflow, sometimes referred to as *interactive machine learning*[34]. This means you can present it some examples of the mapping you want, and it will learn the configuration. I will talk more about interactive machine learning when we start work on the meta-controller; let's get back to the topic of linear regression with neural networks.

10.3 Neural networks and linear regression

Now that you understand what linear regression is (estimating the slope and intercept point of one or more lines) and how it can be used for synthesizer control (mapping between parameters), it is time to learn how to carry out linear regression using a neural network. Those familiar with statistical analysis using languages such as R and Python might think that a neural network is an overly sophisticated tool for this task. There are a couple of reasons for using neural networks though: 1) they provide a convenient workflow for users to iteratively specify a mapping, and 2) unlike more straightforward methods, they can scale to the more demanding task of many-to-many non-linear mapping that we might want to carry out later, with a very similar workflow. The scalability has been one of the drivers of success for neural networks.

10.4 Feed-forward neural networks

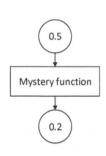

FIGURE 10.4
A neural network applies a function to its input.

A common type of neural network is a feed-forward neural network. Numbers go in one end, and numbers come out of the other in response. From the outside, it is just like a function in C++ where the inputs are the function parameters and the outputs are the return value. Unlike a regular function such as those you might write in C++, a neural network starts off not knowing what exact computations it needs to do to get from the input to the output. It must *learn* what the computation is. Learning functions is what machine learning is all about. But we should not get ahead of ourselves. Before we can consider how a neural network might learn, we need to think about what it is made of and how that makeup allows it to calculate the outputs based on the inputs. Consider the simple feed-forward neural network shown in figure 10.4.

What is the mysterious process? It turns out that in simple feed-forward neural networks, the process is just the same as the definition of a straight line, in other words, the output y is the input x scaled and summed with another value ($y = mx + b$). The difference is that we change the names of the variables. The scaling value m is called a *weight* and the adding value is called a *bias*. So $y = wx + b$.

Given an input of 0.2, what should the weight w and the bias b be to produce an output of 0.5? Can you come up with a solution? You probably realise there are many solutions. I came up with $w = 1$ and $b = 0.3$. That works, but what if we have a second example of an input-output pair, such as 0.4 and 0.9? Do my values for w and b still work? Do yours?

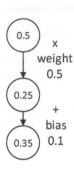

What kind of process did you follow to figure out the weight and bias parameters? In neural networks, the process is referred to as learning and it involves iteratively adjusting the weight and the bias until the network generates correct or close to correct outputs when presented with the inputs. When I told you about the second input-output pair, you may have changed your estimated weight and bias. If you did, you were acting like a neural network! Of course, real neural networks have many weights and biases, allowing them to model sophisticated functions. The weights and biases are referred to as parameters. Our network from figure 10.5 had two parameters. For some perspective, consider that we count the number of parameters for large language model neural networks such as the GPT series in the *billions* or *trillions*. But they still fundamentally take an input and process it in some way to produce an output.

FIGURE 10.5
A neural network scales by a weight and adds a bias.

10.5 Progress check

You should now understand what 1D linear regression is and how it can be expanded to higher dimensions. You should also understand the connection between simple feedfoward neural network units with weights and biases and linear regression.

11

Experiment with regression and libtorch

Now you have some feel for the way a neural network might model a linear relationship between inputs and outputs, it is time to dive into an implementation of this using libtorch. This chapter provides instructions for implementing linear regression with a simple neural network using libtorch. It assumes that the development environment is set up, including an IDE, JUCE, CMake and libtorch. The implementation involves creating a new project, generating a dataset, adding noise to the dataset, creating a neural network model, converting the dataset to tensors, computing the error, learning from the error using the optimiser, and running the training loop. So you are going to be busy. The result of working through this will be a trained neural network that can predict y values from x values with a low error rate. The instructions also include challenges to extend the implementation to higher-dimensional datasets and non-linear mappings.

11.1 Create a new project

Start by creating a new project folder. You can re-use the CMakeLists file from your minimal libtorch project, which was based on repo-guide 39.2.10. You should have a folder structure as follows:

```
1  .
2  |-- CMakeLists.txt
3  '-- src
4      '-- main.cpp
```

Your main.cpp file should be the same as the minimal libtorch example:

```
1  // I have added some extra includes that we will need later
2  #include <vector>
3  #include <iostream>
4  #include <random>
5  #include <torch/torch.h>
6
7  int main() {
8      torch::Tensor tensor = torch::rand({2, 3});
```

```
 9    std::cout << tensor << std::endl;
10  }
```

Note the extra includes I added at the top. We will need those shortly. I will not repeat the CMakeLists.txt file here for brevity, but make sure you have the extra lines for libtorch as seen in listing 39.2.10 in the repo guide.

11.2   Create a dataset

First, you need to create a dataset containing x,y coordinates. You will use this to train and test the neural network. The dataset should be suitable for line fitting via linear regression. Therefore you need to generate the points along a line with some random noise added. Some points should fall above and some below the line. The dataset-generating function should allow the line to be offset (or biased) such that it crosses the y-axis at a certain point and allow the y values to be a consistent multiple of the x values. Let's break that into steps. Step 1 is to generate a linear ramp of values. Here is a function signature for that:

```
 1  /**
 2   * @brief Generate x,y values on a linear ramp. y = m(x) + b
 3   *
 4   * @param b: bias
 5   * @param m: multiplier
 6   * @param startX: start of the value range
 7   * @param end: end of the value range
 8   * @param count : how many x,y pairs to generate
 9   */
10  std::vector<std::pair<float,float>>  getLine(float b, float m,
11                        float startX, float endX, float count)
```

$std::vector$ is a growable array. $std::pair$ is a pair of values. $std::vector < std::pair < float, float >>$ specifies a growable array containing pairs of floats. In other words, our x,y values.

Now the implementation:

```
 1  float y{0}, dX{0};
 2  std::vector<std::pair<float,float>> xys;
 3  dX = (endX - startX) / count; // change in x
 4  for (float x = startX; x < endX; x+=dX){
 5      y = (m * x) + b;
 6      xys.push_back({x,y});
 7  }
 8  return xys;
```

Read it carefully. Can you write some tests to verify that it generates the numbers as expected? Here is an example of a test wherein I ask for 10 values with x in the range of 0 to 10. Notice how I access the first x value with $xys[0].first$. To access the y value, you can use $xys[0].second$. This is because each item in the array is a std::pair, and we access the values in the pair with '.first' and '.second'. So to the test – if the first value is not 0, print an error:

```
std::vector<std::pair<float,float>> xys;
xys = getLine(5.0, 0.5, 0.0, 10.0, 10.0);

if (xys[0].first != 0.0) {
    std::cout << "start value was not zero as expected "<< std::endl;
}
```

I like to use informal unit testing when developing number processing programs. You will see more of that as we work through it.

11.2.1 Dataset experiments

There are some extensions you can try out here. Can you test if the function works correctly if you try to create a ramp that goes down instead of up? In other words, set m to a negative value. Can you write a function that prints the x and y values from the function? Here is a signature:

```
void printXYs(std::vector<std::pair<float,float>>& xys)
```

11.3 Add noise to the dataset

Now that we have a linear ramp of values, we need to make it more challenging for the neural network by adding noise. Here is a function signature:

```
void addNoiseToYValues(std::vector<std::pair<float,float>>& xys,
                       float low, float high)
```

For less experienced C++ programmers, note that we are sending in the 'xys' parameter, which is a vector of pairs. Also, I added the ampersand character & after the type to use 'pass by reference'. This means that C++ will pass in a reference to the xy values instead of making a copy and passing that. Using the ampersand avoids the costly copy operation, but we must remember that the function now has direct access to the original version of the data sent to it.

Here is an implementation using the random functions available in the standard library:

```
std::default_random_engine generator;
std::uniform_real_distribution<float> distribution(low, high);
auto rand = std::bind ( distribution, generator );
```

```
4 for (int i=0;i<xys.size();++i){
5     float r = rand();
6     std::cout << "rand: " << r << std::endl;
7     xys[i].second += r;
8 }
```

Creating a new random number generator every time the function gets called is probably inefficient, but it keeps things simple for now. Try your function out in combination with the printXYs function that you hopefully wrote earlier:

```
1 int main()
2 {
3     std::vector<std::pair<float,float>> xys;
4     xys = getLine(5.0, 0.5, 0.0, 10.0, 5);
5     printXYs(xys);
6     addNoiseToYValues(xys, -0.1, 0.1);
7     printXYs(xys);
8 }
```

I see output like this with my version of the printXYs function:

```
1  x: 0, y: 5
2  x: 2, y: 6
3  x: 4, y: 7
4  x: 6, y: 8
5  x: 8, y: 9
6  x: 0, y: 4.9
7  x: 2, y: 5.92631
8  x: 4, y: 7.05112
9  x: 6, y: 7.99173
10 x: 8, y: 9.00655
```

11.4 Progress check

At this point, you should be able to generate linear ramps of values in specified ranges with a specified number of x,y pairs. You should be able to add noise to the x,y pairs and print them out.

11.5 Finding the best-fit line using a neural network

Now you have a dataset generator, it is time to use libtorch to fit the line. There are a few parts to this:

1. The neural network model that will learn to predict y values from x values.

2. Code to convert our x,y data into tensors so the network can understand it.

3. The error metric that will compute the error over the dataset.

4. The optimiser that will use the computed error to update the neural network weight and bias.

5. The training loop that will iteratively compute and learn from errors.

11.6 Neural network model

First, the neural network model. Neural networks are organised into layers which contain nodes or 'neurons'. We only need a single node for 1D linear regression:

```
auto net = torch::nn::Linear(1, 1);
```

Yes – that is all. Torch's 'Linear' is a simple 'feed-forward' neural network layer. We have sent the Linear function the arguments 1 and 1, which means the layer has one input and one output. Figure 11.1 shows a 1-in, 1-out Linear layer.

According to the documentation for Linear[1], it "Applies a linear transformation to the incoming data: $y = xA^T + b$". Aside from the T component, that expression should be pretty familiar – it is the same as the definition of a straight line, $y = mx + b$. A quick note about the libtorch documentation: the best strategy is to look things up in the Python PyTorch documentation as it has more detail and explanation than the libtorch C++ documentation. Once you have found and read about the functions you want in the PyTorch documentation, you can locate them in the libtorch documentation, as sometimes the names change for C++.

So you have created your first neural network — but what are its parameters? You will be using machine learning to learn these parameters.

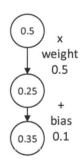

FIGURE 11.1
Simple single layer network with one input and one output.

---

[1] https://pytorch.org/docs/stable/generated/torch.nn.Linear.html

You can print the parameters out as follows:

```
int main() {
    auto net = torch::nn::Linear(1, 3) ;
    for (const auto& p : net->parameters()) {
        std::cout << p << std::endl;
    }
}
```

On my system, this code outputs the following:

```
parameter:   0.3090
[ CPUFloatType{1,1} ]
parameter:  -0.2448
[ CPUFloatType{1} ]
```

If you are working on Windows, you must ensure you are building in the same mode as the libtorch library you are using. If you are using the Debug version of libtorch, build in Debug mode, and the same for Release mode.

So we have two parameters – 0.3090 and -0.2448. Earlier I said that a single 'node' in a feed-forward neural network scales the input by a weight w then adds a bias b. Let's pass some input to the network to verify that is indeed what it is doing. The following code passes a single value through the neural network by calling it as if it were a function. Add the code after your parameter printing code:

```
auto input = torch::empty({1, 1});
input[0][0] = 0.5;
auto output = net(input);
std::cout << "Passing 0.5 in ... this came out:\n"
          << output << std::endl;
```

You might be wondering what empty is and what the data type of 'input' is. You cannot pass simple floats directly into a libtorch neural network – you have to wrap the floats inside a data structure called a tensor. Empty allows you to create an empty tensor of a certain shape (1x1 in this case). Then you can put data into the tensor using array-like syntax – check line 2. More on tensors in a bit but let's just see what comes out of the other end of the network. On my machine I see output like this, noting that the parameters are different every time I run it:

```
0.8999
[ CPUFloatType{1,1} ]
0.3680
[ CPUFloatType{1} ]
Passing 0.5 in ...
0.8179
[ CPUFloatType{1,1} ]
```

Let's verify we know what is going on. The input x was 0.5; the output y was 0.8179; the weight parameter w was 0.8999 and the bias parameter b was 0.3680. Use a calculator to confirm that $y = wx + b$.

11.6.1 More complex model

Just out of interest, let's see what a more complex neural network's parameters look like. Try changing the arguments you sent to Linear to be 2 and 3, to create a network layer like the one shown in figure 11.2. You will also need to change the code that creates the input. Here is the updated code to generate the input – note that the input tensor is now 1×2 in size:

```
1  auto input = torch::empty({1, 2});
2  input[0][0] = 0.5;
3  input[0][1] = 0.25;
```

On my machine I see the following output for the program:

```
1  0.0610   0.5951
2  -0.2292  -0.1606
3  -0.2962  0.5072
4  [ CPUFloatType{3,2} ]
5  -0.1701
6       0.4497
7       0.5419
8  [ CPUFloatType{3} ]
9  Passing 0.5 in ... this came out:
10      0.0092  0.2950  0.5206
11  [ CPUFloatType{1,3} ]
```

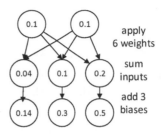

FIGURE 11.2
More complex single layer with more inputs and outputs. Now we apply a weight to each input as it goes to each output. We then sum the weighted inputs and apply a bias to each output.

The first iteration of the parameter printing loop prints six numbers. This is because each of the two inputs is connected to each of the three outputs, requiring six weights. Six connections, six weights. Each of those three outputs will sum its weighted inputs. The second iteration prints a further three numbers. Those are the biases applied to the outputs.

11.7 Converting the dataset to tensors

As you saw just now, to feed some data into the network, you must convert the data from float values into tensors. But what is a tensor anyway? Tensors are the life blood of neural networks, flowing through the network nourishing it with the data they carry. Perhaps that was a bit too colourful. The PyTorch documentation provides a succinct technical description of a Tensor[2]: "A torch.Tensor is a multi-

---

[2]https://pytorch.org/docs/stable/tensors.html

dimensional matrix containing elements of a single data type.". If you were to put data types in order of increasing complexity, you would start with a single float, then a vector of floats, then a matrix and finally, a tensor.

The following lines create a tensor suitable for storing your vector of x,y pairs:

```
torch::Tensor in_t;
in_t = torch::empty({(long) xys.size(), 1});
```

Note that I specified the data type this time, instead of using auto. That makes it clear we are working with a Tensor variable. Then copy the x coordinates into the input tensor:

```
for (int i=0; i<xys.size(); ++i){
    in_t[i][0] = xys[i].first; // first is x
}
```

Now get ready to be impressed by how flexible neural networks are – we can pass the entire input dataset of x values into the network in one go, and it will compute the entire output dataset of its estimated y values:

```
torch::Tensor out_t = net(in_t);
std::cout << "output: " << out_t << std::endl;
```

We call our 'net' variable as if it is a function, passing it the input tensor. 'net' returns another tensor containing the network's guesses as to what the outputs should be. Here is an example of the output I see from the above code:

```
output: -0.7382
0.6678
2.0737
3.4796
4.8856
[ CPUFloatType{5,1} ]
```

Notice that the shape (5 by 1) and type (CPUFloat) of the tensor and the data it contains are printed.

11.8 Computing the error

Now that we have the network output, we can compute the error. We are going to use mean squared error (MSE). Remember that the error is the difference between the output of the network and the output we want. The output we want is the correct set of y values. The error function needs the net output and the correct output as arguments, so we must convert the correct outputs into another tensor.

```
torch::Tensor correct_t;
correct_t = torch::empty({(long) xys.size(), 1});
```

```
3 for (int i=0; i<xys.size(); ++i){
4     correct_t[i][0] = xys[i].second; // second is y
5 }
```

Then compute the error:

```
1 torch::Tensor out_t = net(in_t);
2 torch::Tensor loss_t = torch::mse_loss(out_t, correct_t);
3 float loss = loss_t.item<float>();
```

torch::mse_loss is how we calculate the loss. Try calculating the loss and then looking at the difference between the correct outputs and the outputs from the network. Can you figure out how MSE is being computed? Clue: MSE stands for mean squared error.

11.9   Learning from the error: the optimiser

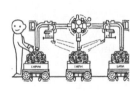

FIGURE 11.3
The optimiser adjusts the network weights.

Learning involves computing the error between the output of the network and the correct output, then adjusting the network's parameters so it can achieve a lower error. To do that, we need something called an optimiser. Various optimisers are available which deal with the learning process in different ways. For example, should the error across the whole dataset be considered or is it good enough just to look at a section of the dataset? How much should the network be adjusted on each iteration? Should the amount it is adjusted change over time, as the network zones in on a good set of parameters? Developing better optimisers, such as the Adam optimiser, was an important element in the rise of deep networks during the 2000s and 2010s.

We will use the stochastic gradient descent (SGD) optimiser. SGD is an optimiser that computes the changes needed to the network from a subset of the input-output pairs in the dataset. So it is quite an efficient optimiser. The following code creates an optimiser:

```
1 torch::optim::SGD optimizer(net->parameters(), lr=0.01);
```

Note how we pass the network parameters to the optimiser when we create it. To use the optimiser to train the network, we can use the following code:

```
1  // compute output
2  out_t = net(in_t);
3  // compute loss
4  torch::Tensor loss_t = torch::mse_loss(out_t, correct_t);
5  // reset the previous network parameter changes
6  optimizer.zero_grad();
7  // compute changes required to network parameters
8  loss_t.backward();
9  // update network parameters
10 optimizer.step();
```

This code represents training for a single epoch. Training for one epoch means you have calculated the error over the entire dataset. With larger datasets, the dataset is split into batches and the network is trained on one batch at a time. This makes it possible to train on larger datasets than you can store in memory. The linear regression dataset is very small so you can process the entire thing in a single batch.

11.10   The training loop

We must run the training process many times to properly train the network, i.e. achieve a low error. A low error means we have found a decent line fit for our dataset's x and y coordinates.

```
1  float loss = 1000;
2  while(loss > 0.5){
3      torch::Tensor loss_t = torch::mse_loss(net(in_t), correct_t);
4      loss = loss_t.item<float>();
5      std::cout << "Loss: " << loss << std::endl;
6      optimizer.zero_grad();
7      loss_t.backward();
8      optimizer.step();
9  }
```

Put this all together, organising it into functions if you wish to. Ensure you pass things like the network and the optimiser to functions 'by reference' (using &) to avoid expensive copy operations. You can find a solution to this in repo-guide 39.3.1. How many epochs does your network take to learn decent values for w and b? How does adding more noise to the data affect the time taken to complete training? How low an error do you think you need? Note that it is not possible to achieve a zero error, i.e. a perfect line that goes through all the data points.

11.11 Training extension: higher dimensions

Can you generate a higher dimensional dataset and an appropriate network to carry out the regression? The example above maps from one-one — can you map two-one or many-many? You might find using the torch functions to generate higher-dimensional datasets easier than coding the dataset generators yourself. For example, check out torch's range[3] and rand[4] functions.

11.12 Progress check

At this point, you should be able to conduct a one-one linear regression using a torch Linear network layer. You should be familiar with datasets, training, errors and epochs.

---

[3] https://pytorch.org/docs/stable/generated/torch.range.html
[4] https://pytorch.org/docs/stable/generated/torch.rand.html

12

The meta-controller

This chapter introduces the AI-music system called the meta-controller which you will be building for the next few chapters. The meta-controller system allows you to control many parameters on one or more instruments and effects simultaneously using a set of powerful 'meta-controls'. The meta-controller is a variant of Rebecca Fiebrink's Wekinator, built using C++ and the libtorch machine learning library; the difference is that the meta-controller is specialised for controlling standard plugin instruments and effects. This chapter describes the original Wekinator system and its workflow, as well as describing some previous variants that people have created such as the learner.js library. At the end of the chapter you should have a clearer idea about Wekinator and the meta-controller's interactive machine-learning workflows.

12.1   What are we going to build?

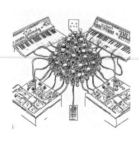

FIGURE 12.1
The meta-controller uses a neural network for new methods of synthesizer sound exploration.

The system we will build is a meta-controller for synthesizers and effects. In a sentence, a meta-controller is a system that allows us to control many parameters on one or more instruments and effects simultaneously using a smaller number of controls. For example, with a meta-controller, you can map positions on a laptop trackpad to presets on a synthesizer, then morph between these presets by moving between positions on the trackpad. The meta-controller can figure out what the synthesizer's parameters should be in between the presets you have mapped to the trackpad. In fact it can infer the synthesis parameters for any arbitrary position on the trackpad.

The meta-controller we create will be able to host native audio plugins. This means anyone using it can control whichever plugins they have installed on their system. The meta-controller will allow you to control more than one plugin at a time, providing new modes of interaction with your plugins and breaking free from the limitations of graphical user interfaces and standard controllers.

12.2   Wekinator: inspiration for the meta-controller

The main inspiration for the meta-controller is Rebecca Fiebrink's 'Wekinator'. According to the Wekinator website[1], 'The Wekinator allows users to build new interactive systems by demonstrating human actions and computer responses, instead of writing programming code'. The concept is that you show Wekinator an input and tell it which output you want when it sees that input again. The clever part is that you can show it several input-output pairs, and after training, it can *infer* what the output should be when you show it an input it has not seen before. Returning to the example of mapping trackpad positions to synthesizer presets, you might place four different presets at the corners of the trackpad. Wekinator will generate and morph between the presets as you move around on the trackpad.

Fiebrink implemented Wekinator as an OpenSoundControl (OSC) message processor. The inputs are OSC messages, and the outputs are more OSC messages. The Wekinator system is flexible and comes with front-ends to convert various types of inputs (sensors, cameras, MIDI control messages, etc.) into OSC messages. It can then send its outputs as controls for any system that receives OSC messages (Ableton Live, SuperCollider etc.). Wekinator is built in Java and uses the Weka machine learning library[18], hence the name.

Since Fiebrink released the original version of Wekinator in 2009, there has been a Wekinator 2.0 and other versions. Louis McCallum worked with Fiebrink to reimplement the Wekinator concept in Javascript using the Tensorflow.js library. McCallum called the new library 'learner.js'[26]. Learner.js works in the web browser and is especially useful for folks wishing to share their creations with others. In fact learner.js was developed for use in the MIMIC[2] browser-based programming environment. MIMIC in combination with learner.js makes sharing and hacking other people's Wekinator examples much easier.

I worked with this team on the MIMIC project and have even performed live with Louis and MIMIC project lead Mick Grierson by collaboratively editing the same Javascript document! As a small anecdote, we agreed to do a live audiovisual performance using the MIMIC browser-based live coding environment during the 2020 Covid-19 lockdown. The performance was part of the concert programme for the NIME conference, which is perhaps the most important academic conference

---

[1]http://www.wekinator.org/
[2]https://mimicproject.com/

for folks developing innovative musical instruments. A few days before our performance, we discovered that our stream would be fed into a real Shanghai club containing real people we could not even see! Luckily nothing crashed too badly despite having three people madly editing the same file in real-time.

Returning to the central theme, we will build a version of Wekinator using C++ and the libtorch machine learning library. Our version will not be as general as Wekinator, and instead, it will be specialised for control of standard plugin instruments and effects. So it will be a plugin host that can learn how to control the plugins in response to different input types.

12.3   Inputs and outputs

Fiebrink describes Wekinator's inputs as 'human actions', and they can be complex things like a facial expression captured via a webcam or simpler things such as an x,y position on a laptop trackpad or a set of slider positions on an on-screen user interface. Whatever the human input, it ends up as an array of numbers inside Wekinator – a frame from a webcam is converted to a list of many RGB or greyscale values, and an x,y position on a trackpad becomes a list of two values.

The user can configure how many numbers Wekinator generates at its output based on their needs. For example, if they have a synthesizer with ten parameters and they want to control it using their trackpad, the input would have two numbers representing x and y on the trackpad, and the output would have ten digits, one for each synthesizer parameter.

12.4   The train-infer-train workflow

The workflow, illustrated in figure 12.2, is a significant innovation in the Wekinator system. It is described as interactive machine learning because the user presents training data to the machine learning algorithm in an interactive manner. Interactive means the user presents training data, trains in real-time, experiments with the trained model and then presents more data if needed. This is different from regular machine learning. In regular machine learning, the training is not interactive, and it generally involves a much larger dataset. Part of the trick here is knowing that you do not need perfect training with a huge dataset to provide a user with a sense of meaningful control. So long as something reasonably control-

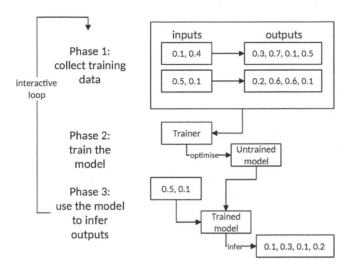

FIGURE 12.2
The Wekinator workflow: data collection, training, inference then back to data collection.

lable happens when the user manipulates the controller inputs, creative work can happen.

12.5   Wekinator's flexibility

Wekinator is a powerful system because of the interactive machine learning workflow and because it is a very general-purpose system. You can feed many types of input data to it – anything you can convert into OSC at any dimensional level. Then you can map the inputs to all kinds and dimensionalities of outputs. One problem with Wekinator is that its workflow is a little tricky to integrate with existing music technology. For example, to control a plugin synthesizer with Wekinator, you would need to figure out how to get your input control data into OSC messages and then convert the output OSC messages into control signals that the DAW hosting the plugin could understand. You would also need to specify how many parameters you wanted to control.

The meta-controller you will develop in the following sections sacrifices a little of the flexibility of Wekinator to gain a deeper integration with standard music technology such as plugins and plugin hosts. But you could adapt it back towards the full Wekinator by adding OSC input and output to it.

12.6 Progress check

You should now be able to explain what the Wekinator is and how it enables its users to use an interactive machine-learning workflow to specify complex mappings between inputs and outputs. You should know that the meta-controller is a variant of the Wekinator which focuses on controlling plugins such as synthesizers and effects.

13

Linear interpolating superknob

In this chapter, you will get started on the meta-controller system. The first step is implementing something similar to the 'superknob' found on Yamaha synthesizers such as the MODX+. The superknob described here allows you to control two synthesizer parameters on a simple FM synthesizer using a single knob. It does so using simple interpolation. This will lay the groundwork for the meta-controller, which allows you to control all parameters on any plugin using a single user interface control.

13.1   FM synthesizer

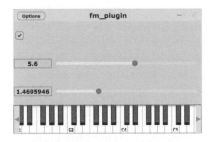

FIGURE 13.1
User interface for the simple two-parameter FM synthesizer. The toggle switch switches between drone and envelope mode, the two sliders control modulation depth and index and the piano keyboard allows you to play notes on the synthesizer.

You should start with the two-parameter FM synthesizer shown in figure 13.1, which can be found in repo-guide 39.2.9. To recap the features of this synthesizer:

1. Two sliders control modulation index and modulation depth

2. A toggle switch enables either constant playing or enveloped playing

3. A piano keyboard allows playing notes on the synthesizer

Make a copy of the project and edit the CMakeLists.txt file so it has your own 'company name' and plugin name/ code. Build and run it in standalone mode to verify it works correctly.

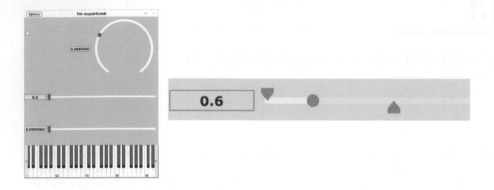

FIGURE 13.2
Superknob UI on the left. On the right is a closer view of a range slider. Small triangles above and below the line allow the user to constrain the range of the main slider control.

13.2 Implement the superknob

Now you have the FM synthesizer up and running, you are ready to implement the basic superknob. First, in the PluginEditor.cpp file, call the setSliderStyle function on the two-parameter sliders. Send it the argument juce::Slider::ThreeValueHorizontal. That will change the sliders from regular, single-position sliders into range sliders allowing you to constrain the value range. Figure 13.2 shows a range slider.

Next, you need to add the superknob to the user interface. This widget is just another slider, but you can call the setSliderStyle function to change it into a rotary 'knob' style. Use the juce::Slider::RotaryHorizontalDrag style. This knob can be adjusted by clicking and holding, then moving left or right, hence 'horizontal drag'. Remember the three-step process to add a widget to your user interface:

1. Add a juce::Slider member to the private section of the class in PluginEditor.h

2. In the constructor in PluginEditor.cpp, call addAndMakeVisible on the new widget and call addListener on the widget

3. In the resized function in PluginEditor.cpp, call setBounds on the widget to place it where you want on the user interface.

I made the superknob quite large for my user interface and placed it at the top of the interface, as shown in figure 13.2. You can design it however you like.

Now you have the superknob widget displayed on the user interface, you need to implement the listener code to respond to it and to cause a change to the two parameters. In the PluginEditor.cpp code, locate the sliderValueChanged function. Add an if statement to check if the incoming slider value is equal to the address of your superknob widget. In my case, the widget's variable is called superKnob, so I have some code like this:

```
if (slider == &superKnob){
//The user adjusted the superknob
}
```

Now your job is to take the value from the superknob widget in the range 0-1 and map it to the range specified by each of the sliders. This is just like the equation for a line: $y = mx + b$, where y is the value for the slider (i.e. modulation index or modulation depth), and x is the value of the superknob widget. To calculate m (also known as the weight), you need to work out the range for the target slider by subtracting its lowest allowed value from its highest permitted value. To calculate b (also known as the bias), just use the getMinValue function:

```
double high = modIndexSlider.getMaxValue();
double low = modIndexSlider.getMinValue();
double weight = high - low; // m
double bias = low; // b
```

getMaxValue tells you where the user placed the high constraint on the three-point slider. getMinValue tells you where the user placed the low constraint. Now, you can plug weight and bias into the equation and send the value to the target slider:

```
double superV = superKnob.getValue(); // x
double newV = (superV * weight) + bias; // y
modIndexSlider.setValue(newV);
```

Repeat the code for the other slider, called modDepthSlider, in the FM synthesizer example code. You should now be able to move the superknob dial and see the two sliders moving in their constrained ranges. You now have your own superknob.

13.3   Progress check

You should be able to explain the concept behind the superknob and how it allows a single knob to control multiple parameters at the same time using range controls amd linear interpolation. You should have an FM synthesizer plugin with a working superknob feature that controls two parameters at the same time in ranges that the user can specify.

14

Untrained torchknob

This chapter provides instructions on replacing the linear interpolation system in the superknob control with a neural network. We shall call this new system the 'torchknob'. The chapter will guide you through creating a simple neural network that can map from a single input control to two output controls, then adding a softmax layer to manage the output ranges. You will create a separate NeuralNetwork class which is designed to provide functions for passing data and training the network. At the end of the chapter, you will be able to control the FM synthesizer controls using a neural network, though not in a very predictable way yet!

14.1 Torchknob software architecture

Figure 14.1 shows a high-level architecture for the torchknob system. This is structurally similar to the superknob from the previous section, but instead of using range controls to specify the mapping from the large dial at the top to the synthesizer parameter sliders at the bottom, the user trains a neural network to learn the mapping. The aim is to implement the Wekinator workflow mentioned previously:

1. Set the parameter sliders to the desired position

2. Set the torchknob control to the desired position

3. Store the torchknob → sliders mapping as a training data point

4. Back to step 1. until the user has defined as many input-output associations as they like

5. Train the neural network on the set of torchknob → sliders mappings

6. Move the torchknob at the top, and the neural network will automatically set the sliders to the desired positions, inferring or regressing between the points in the training data

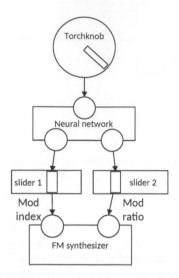

FIGURE 14.1
The torchknob system architecture.

14.2 Quick hack to pass data through a network

Start by making a copy of your FM superknob project. Remember to rename the plugin and give it a unique code so the new project does not clash with the other one in your plugin lists. Once you have a copy of your SuperKnob project, edit your CMakeLists.txt file to include libtorch in the project. You can find instructions on how to do this in section 5.9.

Next you can try a quick hack to pass data from one slider via a neural network into the other two sliders. In PluginEditor.h, make sure you include the torch/torch.h header. Then add this to the private section of the editor class:

```
torch::nn::Linear linear{1, 2};
```

That creates a class member with the type torch::nn::Linear and the shape: '1 input, two outputs'. Just what we need to map from one slider to two other sliders. Then you can use this code to pass the superknob value through the linear layer in the sliderChanged function in PluginEditor.cpp:

```
 1  // create a 1x1 tensor
 2  torch::Tensor in = torch::empty({1, 1});
 3  //Copy the superknob value into the tensor
 4  in[0][0] = superKnob.getValue();
 5  //Pass the tensor through the linear layer
 6  torch::Tensor out = linear(in);
 7  // print the result (DBG does not work)
 8  std::cout << out << std::endl;
 9  //Extract the result from our tensor
10  // with out[0][1].item<float>()
11  // which specifies the position and
12  // data type you want to convert the tensor
13  // data into
14  modDepthSlider.setValue(out[0][0].item<float>());
15  modIndexSlider.setValue(out[0][1].item<float>());
```

Try moving the superknob around. You should see the two synthesis parameter sliders moving. I say 'should' because the network starts off with random weights and biases, so the two sliders might not move much or at all because the values coming out of the network are out of range of the slider. Each time you run the program, the network weights and biases (parameters) will be different. Keep re-running until you get some movement.

14.3   Basic NeuralNetwork class

Now you have a proof of concept in place; you can read values from a slider, feed them through a neural network, read the outputs and use them to set values on two other sliders. The problem is that the code will become messy once you progress onto more complex networks and once you want to start training. You will have a mix of user interface and neural network code in your PluginEditor files. The solution to this problem is to begin work on a separate NeuralNetwork class that provides a simple set of public functions for passing data in and training. When designing classes, I usually start with an 'aspirational' header file containing everything I wish I had (aside from a Lamborghini, of course). For example, in the header file, I specify the functions I would like to have. So let's do that. Create two new files in your src folder: NeuralNetwork.h and NeuralNetwork.cpp. In NeuralNetwork.h, put the following 'aspirational' class definition:

```
1  #pragma once
2
3  #include <torch/torch.h>
4
5  class NeuralNetwork : torch::nn::Module {
6      public:
7          NeuralNetwork(int64_t n_inputs, int64_t n_outputs);
8          std::vector<float> forward(const std::vector<float>& inputs);
9          void addTrainingData(
10             std::vector<float>inputs,
11             std::vector<float>outputs);
12         void runTraining(int epochs);
13     private:
14         int64_t n_inputs;
15         int64_t n_outputs;
16         torch::nn::Linear linear{nullptr};
17         torch::Tensor forward(const torch::Tensor& input);
18
19 };
```

Some comments on that code:

1. The NeuralNetwork class inherits from torch::nn::Module. This means it inherits all the functions of that parent class. Inheriting from Module is the recommended way to create a network in Torch.

2. The NeuralNetwork constructor takes two integers as arguments, specifying the number of inputs and the number of outputs.

3. The public forward function takes a vector of floats as input and provides a vector of floats at its output. Whilst the network itself will need the data in the form of tensors, from the JUCE side, it is cleaner if we work in floats and vectors as the JUCE code then does not need to know anything about libtorch.

4. There is a private function called forward, which receives and returns tensors. The other forward function will use this function internally to complete the inference.

5. Both forward functions specify that their arguments are const references (&). For those who are a bit new to C++, this means we avoid copying the incoming data every time the functions are called (pass by reference), and we are not allowed to change that incoming data (const).

6. There is an addTrainingData function which will enable us to add training examples (input-output pairs).

7. There is a runTraining function which will eventually allow us to train the network against the training examples.

8. There is a private class member called linear, a Linear network layer. It starts being set to a nullptr value.

You should now write comments above each function to explain what it does. Having written your documentation, it is time to turn your attention to the NeuralNetwork.cpp file. This is where we make good on the promises in the header file. You need to add NeuralNetwork.cpp to the CMakeLists.txt file in the target_sources command so that the new class is included in the build. Here is an example of a CMake target_sources command which adds the NeuralNetwork.cpp file to the build for a target called fm-torchknob:

```
target_sources(fm-torchknob
    PRIVATE
    src/PluginEditor.cpp
    src/PluginProcessor.cpp
    src/NeuralNetwork.cpp)
```

You might add the header files to the list as well, if that is appropriate for your setup. I have seen people adding header files when working in Visual Studio Community as it makes them easier to find in the project. Now back to the NeuralNetwork.cpp file, here is an implementation for the NeuralNetwork constructor which takes two 64-bit integers for the input and output size and uses them to initialise a Linear layer:

```
NeuralNetwork::NeuralNetwork(int64_t _n_inputs, int64_t _n_outputs)
: n_inputs{_n_inputs}, n_outputs{_n_outputs}
{
    linear = register_module(
            "linear",
            torch::nn::Linear(n_inputs, n_outputs)
        );
}
```

You can see here that I am calling register_module, which is a function inherited from the torch Module class. This allows the Module parent class to keep track of the layers (modules) we are adding to the network. Later you might use the functions provided by Module to do various useful operations on the network as a whole, which is why you need to register the modules.

Now the rest of the functions – here is some placeholder code, for now, just to get it to compile:

```
std::vector<float> NeuralNetwork::forward(
            const std::vector<float>& inputs)
{
    std::vector<float> out = {0, 0};
    return out;
}

void NeuralNetwork::addTrainingData(
            std::vector<float>inputs,
            std::vector<float>outputs)
{
```

```
13 }
14
15 void NeuralNetwork::runTraining(int epochs)
16 {
17
18 }
19
20 torch::Tensor NeuralNetwork::forward(const torch::Tensor& input)
21 {
22     torch::Tensor out = linear(input);
23     return out;
24 }
```

To make sure compiling and linking are working, add a NeuralNetwork to the private area of the PluginEditor.h file:

```
1 NeuralNetwork nn{1, 2};
```

Then in the sliderValueChanged function, put this in the block that responds to the superknob:

```
1 // call forward on NeuralNetwork
2 std::vector<float> nn_outs = nn.forward(
3     // pass it a vector containing the superknob value
4     std::vector<float>{(float)superKnob.getValue()});
5
6 // use the return data to set the slider values
7 modDepthSlider.setValue(nn_outs[0]);
8 modIndexSlider.setValue(nn_outs[1]);
```

This code calls forward on our new NeuralNetwork class, passing it the superknob value and uses the output to set the values on the other two sliders. Build and run to check for any mistakes. Note that the public forward function on NeuralNetwork currently returns two zeroes – it does not actually use the neural network model to infer the slider values.

14.4   Unit testing

At this point, it might be worth setting up your CMake project to allow for lightweight unit testing. I do not know how fast your compile time is, but building a complete JUCE application every time I tweak the neural network code is slower than I would like on my system. You can create a simple program to test your NeuralNetwork class which will build and run faster, allowing for a faster development cycle. There are various formal ways to conduct unit testing with CMake, but here is my simple and informal method.

Add a new file to your src folder called test_nn.cpp. Put the following in that file:

```
1 #include "NeuralNetwork.h"
2 #include <iostream>
3
4 int main()
5 {
6     NeuralNetwork nn{2, 2};
7     std::cout << nn.forward({0.1, 0.5}) << std::endl;
8     return 0;
9 }
```

As you can see, we are just creating a NeuralNetwork object and then calling forward on it, passing in some values. To add this to your CMake project, add the following lines to the end of CMakeLists.txt:

```
1 add_executable(test_nn
2   src/NeuralNetwork.cpp
3   src/test_nn.cpp)
4 target_link_libraries(test_nn "${TORCH_LIBRARIES}")
5 set_property(TARGET test_nn PROPERTY CXX_STANDARD 14)
```

When you re-run CMake to regenerate the project, you should have another executable target you can build. In VSCode, this is accessible from the dropdown menus for the build and run buttons. You can now choose if you want to do quick tests of the NeuralNetwork class by editing test_nn.cpp or if you want to test the full integration with the FM synthesizer. You can add a series of test functions to the test_nn.cpp file.

14.5   Implement NeuralNetwork's forward function

Next, you will replace the code in the NeuralNetwork.cpp file for the two forward functions with some proper code to pass data through the network. First, the public function:

```
std::vector<float> NeuralNetwork::forward(std::vector<float> inputs)
{
    // copy input data into a tensor
    torch::Tensor in_t = torch::empty({1, n_inputs});
    for (long i=0; i<n_inputs; ++i){
        in_t[0][i] = inputs[i];
    }
    // pass through the network:
    torch::Tensor out_t = forward(in_t);
    // initialise size to n_outputs
    std::vector<float> outputs(n_outputs);
    // copy output back out to a vector
    for (long i=0; i<n_outputs; ++i){
        outputs[i] = out_t[0][i].item<float>();
    }
    return outputs;
}
```

Some comments on the forward function:

1. Essentially, it makes a tensor for the input data, copies the vector data to the tensor, passes the tensor to the network, copies the output of the network back out to a vector and returns it.

2. The code is generalised, so it can deal with any number of inputs and any number of outputs.

3. It uses some dynamic memory allocation to create the tensors and vectors. This is definitely not the most efficient way to implement this code but this code will run at 'GUI' speed as opposed to audio speed so it does not need to be highly optimised yet.

4. A more efficient way would be to create the vectors and tensors as class data members and re-use them, but that would complicate the code, and I want to keep it as clear as possible for now.

Next, the private forward function:

```
torch::Tensor NeuralNetwork::forward(const torch::Tensor& input)
{
    torch::Tensor out = linear(input);
    return out;
}
```

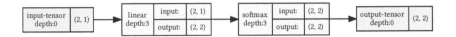

FIGURE 14.2
Basic architecture where a linear layer passes into a softmax layer. The numbers in the brackets indicate input and output shape. The linear layer input (2,1) goes from 1 value to 2 nodes, then output (2,2) goes from 2 nodes to 2 outputs.

Now try re-running the test_nn program a few times. You should see different outputs each time. Can you figure out why the outputs differ each time? Are these outputs in suitable ranges for the sliders?

14.6 Normalising the network output: softmax

The neural network outputs can be in all kinds of ranges, depending on the weights and biases in the network and the input. To push the outputs into more predictable ranges, libtorch provides a few options and softmax is one such option. Softmax takes a set of output values and modifies them so they add up to a total of 1. For the mathematically minded, it can be expressed as follows:

$$Softmax(x_i) = \frac{exp(x_i)}{\Sigma_j exp(x_j)} \tag{14.1}$$

In other words, the softmax value of output x_i is the exponential of that output divided by the sum of the exponents of all outputs. Let's run through a calculation to make it really clear. Say we had two outputs with values 10.2 and 1.5. We take the exponent of those values:

$$exp(10.2) = 26903.1861 \tag{14.2}$$

$$exp(1.5) = 4.4817 \tag{14.3}$$

We sum those values:

$$exp(10.2) + exp(1.5) = 26907.6678 \tag{14.4}$$

Then to compute softmax of 10.2:

$$\frac{exp(10.2)}{exp(10.2) + exp(1.5)} = 0.9998 \qquad (14.5)$$

and softmax of 1.5:

$$\frac{exp(1.5)}{exp(10.2) + exp(1.5)} = 0.0002 \qquad (14.6)$$

Since the sum of the softmax of all outputs is 1, softmax is typically used to convert output layer values into probabilities. This is useful for classification, which is a common task for neural networks. For example, does a given image contain a dog or a cat? In classification tasks, the outputs represent the network's estimation of the probability of an input belonging to a given class, and softmax is a handy layer that converts outputs in any range to outputs in the range 0-1, totalling 1. So you can use softmax as a convenient processing layer to put the outputs in predictable ranges. With the NeuralNetwork class, it is fairly easy to add new layers. Here are the steps to add a new layer:

Step 1: Add a field of the appropriate type for your layer to the private section of the NeuralNetwork class in NeuralNetwork.h. For a softmax layer:

```
torch::nn::Softmax{nullptr};
```

Step 2: Call register_module in the constructor of NeuralNetwork:

```
1  softmax = register_module(
2    "softmax", // name it
3    torch::nn::Softmax(1) // configure it
4  );
```

Note that softmax takes one argument to its constructor, and a '1' is appropriate for our purposes. According to the PyTorch documentation, this constructor argument 'dim' is 'A dimension along which softmax will be computed (so every slice along dim will sum to 1).'

Step 3: Pass the output of the linear layer through the softmax layer in the NeuralNetwork's private forward function. Here is the new, complete forward function:

```
1  torch::Tensor NeuralNetwork::forward(const torch::Tensor& input)
2  {
3      std::cout << "forward input " << input << std::endl;
4      torch::Tensor out = linear(input);
5      std::cout << "forward after linear " << out << std::endl;
6      out = softmax(out);
7      std::cout << "forward after softmax " << out << std::endl;
8      return out;
9  }
```

Note that I have added some print statements to the code so we can check how the data flows through the network. You can see a visualization of the neural network architecture in figure 14.2.

14.7   Testing the simple model

Hopefully, you followed the instructions earlier and set up a simple test program as an additional executable in your CMake project. If you did, you could quickly try out the following test program to check on the output of your neural network:

```
#include "NeuralNetwork.h"
#include <iostream>

int main()
{
    NeuralNetwork nn{1, 2};
    for (float i=0;i<1; ++i){
        std::cout << nn.forward({i/10}) << std::endl;
    }
    return 0;
}
```

On my system, I see output as follows (slightly truncated):

```
forward input   0
[ CPUFloatType{1,1} ]
forward after linear   0.2885   0.4512
[ CPUFloatType{1,2} ]
forward after softmax   0.4594   0.5406
[ CPUFloatType{1,2} ]
0.459409 0.540591
...
forward input   0.4000
[ CPUFloatType{1,1} ]
forward after linear   0.5363   0.1527
[ CPUFloatType{1,2} ]
forward after softmax   0.5947   0.4053
[ CPUFloatType{1,2} ]
0.594736 0.405264
----
forward input   0.9000
[ CPUFloatType{1,1} ]
forward after linear   0.8460  -0.2205
[ CPUFloatType{1,2} ]
forward after softmax   0.7439   0.2561
[ CPUFloatType{1,2} ]
0.743922 0.256078
```

You can now return to your JUCE synthesizer and test the new neural network. Remember that the network outputs are in the range [0..1], so you should scale them into the correct ranges for the sliders.

14.8   Progress check

At this point, you should have a working JUCE/ libtorch plugin project. You should have a neural network class and be able to take data from a JUCE slider and pass it through the neural network. You should be able to pass the output of a linear layer in the network through a softmax layer to normalise it. You should be able to use the normalised neural network output to move two other sliders. You should have a test program that appears as a separate target in your CMake project.

15

Training the torchknob

The chapter begins by explaining how to train the neural network that powers the torchknob. First, the focus is on preparing the user interface for training by adding data input knobs and function buttons. Then, the process of gathering a training dataset by capturing input and output values from the user interface is demonstrated. After that, the chapter describes how the training data is converted to tensor format and stored. The implementation of the training loop is then discussed, wherein a Stochastic Gradient Descent optimiser is used to update the neural network parameters. Additionally, the chapter reconsiders using the softmax layer for output normalization. It suggests replacing it with a sigmoid activation function, the output normalization method utilised in the original Wekinator and learner.js systems.

15.1   Preparing the user interface for training

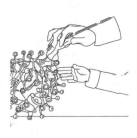

FIGURE 15.1
Interactive machine learning provides more intuitive training for neural networks.

You need a dataset containing example input and output pairs to train a neural network. The idea is that you will use this data to train the network to produce the correct output when presented with the correct input. Once it has learned the mapping it can infer new outputs given inputs it has not necessarily seen before. To put this in the context of the FM synthesizer, you are saying: 'When the superknob is here, I want the other sliders to be here'. You usually need several examples to train the network. You will probably be aware that giant datasets containing millions or billions of examples are needed to train large networks. For our purposes, though, we can achieve decent functionality with just a few examples. For those more versed in neural networks, we will exploit over-fitting as we are just looking for interesting regressions between the data points.

We will gather the training data using an iterated 'learn

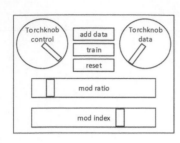 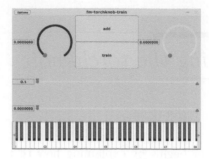

FIGURE 15.2
User interface mockup for trainable superknob system (left). We have an additional knob to specify training input without triggering the movement of the sliders. Actual user interface prototype (right).

by example' approach, sometimes referred to as *interactive* machine learning. We need to change the user interface to have the required elements to allow the user to interactively present examples to the training system. Figure 15.2 shows a wireframe mockup of the interface. It has the following additional elements:

1. A training knob to allow you to specify the inputs for training without the sliders moving

2. An add button to add a training example to the training dataset

3. A train button to trigger training

Your task is to implement the code necessary in the PluginEditor class, so it has the user interface elements shown in 15.2. Go ahead and add the UI widgets to your PluginEditor.h and cpp files. Ensure you add the PluginEditor as a listener to the buttons and sliders. You will implement the actual listener code in the next section. You do not need to stick to the exact design and layout shown – there are other ways to implement the UI. You could work out a way to have a single knob and some sort of 'training' mode. Then you would have a toggle to switch between train and infer mode. That requires somewhat more complex application logic, so I leave it to you to decide if you want to implement that. If you want to change the colours of the widgets, here is some example code wherein I change the colour of a UI element called superKnobTrain:

```
superKnobTrain.setColour(
  // which part of the slider are we changing?
  juce::Slider::ColourIds::rotarySliderOutlineColourId,
  // what is the new colour?
  juce::Colours::aliceblue
);
```

15.2 Gather a dataset

Now you have the user interface up and running, you can implement the code to add training examples for the neural network. You need to gather data from the user interface and then pass it to the NeuralNetwork class, which stores it. Locate the buttonClicked function in PluginEditor.cpp. Write some code that responds to the add button. In my case, the add button is called 'addExampleButton':

```
if (btn == &addExampleButton){
    // read the value from the training control:
    float in = (float) superKnobTrain.getValue();
    // read the current values for the synth parameter sliders:
    float out1 = (float)modDepthSlider.getValue() /
                        modDepthSlider.getMaximum();
    float out2 = (float)modIndexSlider.getValue() /
                        modIndexSlider.getMaximum();
    // send the input-output pair to the neural network
    // so it can be used in training later
    nn.addTrainingData({in},{out1, out2});
}
```

15.3 Implement training data storage

Now you have completed the UI side, it is time to implement the training data storage in the NeuralNetwork class. Add these data members to the private section of NeuralNetwork.h:

```
std::vector<torch::Tensor> trainInputs;
std::vector<torch::Tensor> trainOutputs;
```

You can see that these variables store vectors of tensors. I will admit that there are more 'torchy' ways to manage a dataset, but it is the simplest for now, without digging into the torch::datasets classes. Then you need to change addTrainingData so it converts the incoming vectors of floats into tensors and stores them into those tensor vectors:

```
// converting the input vector 'inputs'
// to a tensor
torch::Tensor inT = torch::from_blob(
    // get access to the raw memory storing
    // the vector of floats
    (float*)(inputs.data()),
    inputs.size()
    // then make a copy of it
```

```
 9      ).clone();
10 // same process for the outputs
11 torch::Tensor outT = torch::from_blob(
12     (float*)(outputs.data()),
13     outputs.size()
14     ).clone();
15 trainInputs.push_back(inT);
16 trainOutputs.push_back(outT);
```

The from-blob function allows you to create a tensor directly from the data in the vector which is a bit more efficient on lines of code and memory allocation than manually coping the vector's data into the tensor. The clone function copies the data from the vector to ensure that data belongs to the trainInputs vectors (otherwise, the memory storing the data can be reassigned). As before, drop down to your test_nn.cpp file to quickly test that things are working as expected. Use printouts to verify the data from the sliders ends up in the training tensors. The test file is handy because JUCE's DBG function cannot handle printing tensors. When I was debugging this code, it helped me to verify data coming into the network.

Another useful technique for testing and managing errors is assertion. To verify that the data being sent to the addTrainingData function has the correct shape, you can add the following two lines to the addTrainingData function in Neural-Network.cpp:

```
1 assert(inputs.size() == (unsigned long) n_inputs);
2 assert(outputs.size() == (unsigned long) n_outputs);
```

That code will trigger a crash if the vectors have the wrong size. Better than limping on and crashing out later with an intractable bug. Note that I had to cast the n_inputs variable from its initial type of int64_t into an unsigned long to avoid a compiler warning. Tensor sizes are int64_t but vector sizes are unsigned longs.

15.4   Write the runTraining code

The final stage is to implement the runTraining function, which will train the neural network using an error metric and an optimiser. The components at play here are similar to those from the linear regressor example you worked on before. They are just reorganised into the NeuralNetwork class. First, the optimiser – add a data member to the private section of NeuralNetwork.h

```
1 // unique ptr so we can initialise it
2 // in the constructor after building the model
3 std::unique_ptr<torch::optim::SGD> optimiser;
```

Note that I am using a unique_ptr here. This allows me to decide which parameters the optimiser receives from its constructor after I have created the neural network model. In fact, I only know what the parameters are once I create the model by calling the register_module function in the neural network's constructor. So I have to dynamically create the SGD. I cannot use this style as I did for the network layers:

```
torch::optim::SGD optimiser{nullptr};
```

SGD, unfortunately, does not have a constructor with that signature. If you are not very experienced with C++, you may not be familiar with unique_ptr or even pointers. What you need to know now is that pointers were inherited by C++ from the C language, and they allow programs to request memory from the operating system when they are running, as opposed to requesting all needed memory at the start. If you use a pointer to allocate memory, you must also return that memory to the system. Otherwise, you have a memory leak, where your program takes memory but does not return it. Pointers are notorious for striking fear and confusion into the hearts of novice C++ programmers, and they can cause all kinds of problems. Unique_ptr and other so-called 'smart pointers' were invented in a later version of C++ to help solve some of these problems. Smart pointers automatically free the memory they use once they go out of scope. In other words, they behave more like regular variables with the added capability of runtime memory allocation. You will see smart pointers such as unique_ptr and shared_ptr popping up in many of the JUCE example applications that come with the JUCE developer kit.

I am using a smart pointer here as it allows me to create an SGD variable without assigning an actual SGD object to it. This is helpful as I can only create an SGD object if I know the parameters for the neural network it will optimise. So with my smart pointer, I can create the object after I have created the neural network model in the NeuralNetwork class constructor. Here is some code that uses the std::make_unique function to create an SGD object wrapped in a unique_ptr. Put the code at the end of the NeuralNetwork constructor in NeuralNetwork.cpp:

```
optimiser = std::make_unique<torch::optim::SGD>(
    this->parameters(), .01);// params, learning rate
```

Note how I call the parameters function on 'this'. Remember that our NeuralNetwork class inherits from the torch Module class. This means it inherits a set of functions from Module. One of those functions is 'parameters', and it returns the parameters of the registered modules in a form that the optimiser can understand. 'this' refers to the current NeuralNetwork object. I could just have called 'parameters' without 'this', but I prefer this more explicit syntax as it tells you parameters is a function inside the current object, not something global or otherwise in the namespace somewhere.

Next, the actual runTraining function. This is where things get somewhat complex as we first convert the training data into a large tensor, then use it in combination with the optimiser to improve the weights:

```
1  //Push inputs to one big tensor
2  torch::Tensor inputs = torch::cat(trainInputs)
3  .reshape({(signed long) trainInputs.size(), trainInputs[0].sizes()
       [0]});
4  //Push outputs to one big tensor
5  torch::Tensor outputs = torch::cat(trainOutputs)
6  .reshape({(signed long) trainOutputs.size(), trainOutputs[0].sizes()
       [0]});
7  //Run the training loop
8  for (int i=0;i<epochs; ++i){
9      //Clear out the optimizer
10     this->optimiser->zero_grad();
11     auto loss_result = torch::mse_loss(forward(inputs), outputs);
12     float loss = loss_result.item<float>();
13     std::cout << "iter: " << i << "loss " << loss << std::endl;
14     // Backward pass
15     loss_result.backward();
16     // Apply gradients
17     this->optimiser->step();
18 }
```

Again, I recommend dropping to your test_nn.cpp file and trying a simple test program. Here is an example of a simple testing program:

```
1  int main()
2  {
3      NeuralNetwork nn{1, 2};
4      for (float i=0;i<10; ++i){
5          nn.addTrainingData({i/10}, {i/5, i/3});
6      }
7      nn.runTraining(10);
8      return 0;
9  }
```

Satisfy yourself that the data is flowing around the system correctly with prints and tests in the test function, then return to the JUCE application and verify that it also works correctly.

15.5 Experiment with training

Figure 15.3 shows an example experiment you can carry out with the torchknob-controlled synthesizer. Try training the network for different numbers of epochs. Can you achieve a satisfactory result where the model moves the sliders to the correct positions you trained it on?

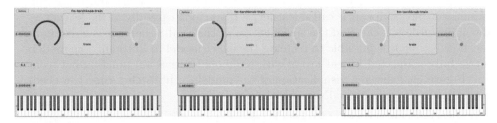

FIGURE 15.3

Example of an experiment you can carry out. First, set the training slider to its lowest value, the same for the modulation controls. Add a training point. Then move to the middle positions, and add a training point. Finally, move to the highest positions, and add a training point.

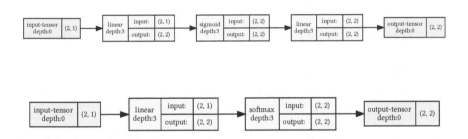

FIGURE 15.4

The learner.js/ Wekinator regression architecture (top). The simpler architecture we used previously (bottom).

15.6   The ideal model

If you carried out the experiment as suggested, you probably found that the basic model with one linear layer followed by a softmax did not manage to learn the mapping between inputs and outputs very well – the torchknob did not move the sliders to the correct positions. Do not lose hope – we just need to adjust the model! Sorry about that – I should have given you the ideal model at the start. But you learned more about training and softmax as a result, which might come in useful later.

I mentioned earlier that Louis McCallum created an implementation of the

Wekinator in Javascript. It is called learner.js. If you look carefully at the learner.js code on github[1], you can find the specification of the model there. It is written in tensorflow.js, but it is possible to recreate the exact model in libtorch. The model has two fully connected Linear layers with a Sigmoid activation function between them. Figure 15.4 shows a visualization of the model. In libtorch terms, you need three variables in your NeuralNetwork class:

```
1 torch::nn::Linear linear1{nullptr};
2 torch::nn::Sigmoid sig1{nullptr};
3 torch::nn::Linear linear2{nullptr};
```

Set them up in the constructor like this:

```
1 linear1 = register_module(
2     "linear1",
3     torch::nn::Linear(n_inputs, n_outputs)
4 );
5 sig1 = register_module("sig1", torch::nn::Sigmoid());
6 linear2 = register_module(
7     "linear2",
8     torch::nn::Linear(n_outputs, n_outputs)
9 );
```

Then process the inputs like this:

```
1 torch::Tensor out = linear1(input);
2 out = sig1(out);
3 out = linear2(out);
```

Note that there is no activation function after the second linear layer, or even a softmax layer. But since it is well tested within the Wekinator systems, this model should be able to learn the mappings much more effectively than the simpler linear/ softmax model can. Run some tests with your test_nn.cpp setup. Does it achieve lower errors than the linear/softmax network? If you want to dig more into the experimental process here, you can create two NeuralNetwork classes, one with the original model and the other with this new one. Compare and contrast using test code. Being able to compare the performance of different neural network models on the same dataset is an important skill to learn.

In order to avoid offending experienced machine learning engineers, I should note that this method of evaluating neural networks by measuring their error on the training set is not the standard way to do things. Typically, you would split the dataset into a training set, a validation set and maybe even a test set. The training is carried out using the training set, you use the validation set error to monitor the progress of training to avoid over-fitting, and you use the test set as an additional resource to validate the training. But here we have a very small training set which might only consist of a few items. So it is not possible to split the data like that. This workflow is designed to allow for simple, interactive machine learning with a

---

[1]https://github.com/Louismac/learnerjs/blob/main/libs/learner.v.0.5.js

small amount of data. In other examples in the book, I will demonstrate the full and proper process of training a neural network with a non-interactive workflow.

15.7   The ideal training loop

Now you have the ideal model for the task, you need a smarter training function. Smarter in that it has a few more measures in there to decide when the training is complete. Right now, the training loop just runs for the specified number of epochs. But how do you know if it is trained enough? One approach would be to observe the change in error after each epoch. If the change becomes very small, it is trained enough. Try this in your training loop:

```
float loss{0}, pLoss{1000}, dLoss{1000};
for (int i=0;i<epochs; ++i){
    //Clear out the optimiser
    this->optimiser->zero_grad();
    auto loss_result = torch::mse_loss(forward(inputs), outputs);
    loss = loss_result.item<float>();
    loss_result.backward();
    this->optimiser->step();
    //Now compute the change in loss
    // and exit if it is too low
    dLoss = pLoss - loss;
    pLoss = loss; // prev loss
    if (i > 0){// only after first iteration
        if (dLoss < 0.00001) {
            std::cout << "dLoss low, exit on ep " << i << std::endl;
            break;
        }
    }
}
```

How many epochs does it take you to train in the test program? My test program takes around 5–6,000 epochs, and the main torchknob program takes about the same number. However, this depends on how many data points you have to feed it. Try adding more layers to the network or increasing the 'hidden layer' size. Adding a reset button that deletes the training data and randomises the network weights would be helpful.

15.8   Next steps

Here are few more experiments you can try:

1. Try passing the NN conflicting values. Observe that it cannot work miracles – if you put two data points in two different places, it will just sit in the middle to minimise the error.

2. Make a more complex synthesis algorithm that uses more parameters.

3. Make a custom or more sophisticated GUI element that generates more control inputs, e.g. a 2D trackpad-type controller.

15.9   Progress check

At this point you should be able to train a neural network to learn a regression through parameter space. You should be able to do so by interactively creating a dataset of input-output pairs using a user interface. You should have experimented with different training loops and different neural network models.

16

Plugin meta-controller

This chapter covers the steps involved in adapting the torchknob FM synthesizer into a plugin host. This means we can move from the torchknob concept, wherein a neural network learns to control a fixed synthesizer's parameters, to the meta-controller concept, wherein a neural network can control any plugin synthesizer available. The steps covered are removing the FM synthesizer UI and DSP code, adding code to load a VST plugin synthesizer, and hooking the plugin into the processBlock function so it can generate audio and process MIDI.

16.1 Hosting plugins

FIGURE 16.1
Hosting plugins allows you to control more advanced synthesizers.

The first step is to adapt the torchknob FM synthesiser plugin into a plugin host. The concept is that your plugin will use an external plugin to generate its audio instead of the FM synthesiser DSP code.

Make a copy of the FM torchknob you created in the previous sections. A reference implementation of this project is described in section 39.3.4. You can start with that reference implementation if you want to follow these instructions as closely as possible. Or, if you want a shortcut, you can skip to section 16.2 and start with the clean project 39.3.5.

Remember to change all occurrences of the project name in the CMakeLists.txt file and to change the PLUGIN_CODE. In my case, working from the FM synthesiser reference project, I carried out a 'find and replace' replacing 'fm-torchknob-train' with 'plugin-host' and then changed the code from cbc3 to dbc1.

Next, you need to enable plugin hosting capabilities in your project. Locate the 'target_compile_definitions' block in the CMakeLists.txt file. Set the property 'JUCE_PLUGINHOST_VST3' to 1, e.g.

```
1  JUCE_PLUGINHOST_VST3=1
```

That will enable plugin hosting capability in your application. Now generate the project, build and run to verify things are ready.

16.1.1   Remove the FM synthesis UI code and DSP

Starting from the reference project, the UI controls to remove are the two sliders for modulation depth and modulation index and the toggle button for the constant envelope mode. My approach here is to comment out the UI objects in the PluginEditor.h file:

```
1  // juce::ToggleButton envToggle;
2  // juce::Slider modIndexSlider;
3  // juce::Slider modDepthSlider;
```

Then in the PluginEditor.cpp, comment out any code that uses those variables, i.e. the relevant lines in the constructor, the resized function, the sliderValueChanged function and the buttonClicked function. Once you have done this, build and run to verify it still builds.

Next, for the code in PluginProcessor. Remove all synthesis-related variables from the private section of PluginProcessor.h except:

```
1  juce::MidiBuffer midiToProcess;
```

Remove all synthesis-related functions in PluginProcessor.h, essentially, anything that is not an overridden function. Leave this one, though:

```
1  void addMidi(juce::MidiMessage msg, int sampleOffset);
```

We will need that to make the piano keyboard widget on the UI work. Now go into PluginProcessor.cpp and remove any lines referring to those variables and definitions of functions you have removed. Your IDE will probably go crazy with red lines and warnings during this process. Finally, you can strip back the processBlock function to something like this:

```
1  // add collected MIDI to the function's MIDI input
2  if (midiToProcess.getNumEvents() > 0){
3      midiMessages.addEvents(midiToProcess, midiToProcess.
       getFirstEventTime(), midiToProcess.getLastEventTime()+1, 0);
4      midiToProcess.clear();
5  }
6  //Just clear the buffers
7  juce::ScopedNoDenormals noDenormals;
8  auto totalNumInputChannels  = getTotalNumInputChannels();
9  auto totalNumOutputChannels = getTotalNumOutputChannels();
10 for (auto i = totalNumInputChannels; i < totalNumOutputChannels; ++i)
11     buffer.clear (i, 0, buffer.getNumSamples());
```

For completeness, I also renamed the two classes: FMPluginProcessor becomes PluginHostProcessor, and FMPluginEditor becomes PluginHostProcessor.

If it all goes terribly wrong, you can refer to the project 39.3.5.

16.2   Load a plugin: hard coded location

At this point, you should have a stripped-back plugin project which includes the
NeuralNetwork class code and the training control widgets.

We shall start by loading a plugin from a hard-coded location on the file system.
Declare a public function in PluginProcessor.h:

```
void loadPlugin(const juce::File& pluginFile);
```

Const tells the compiler we will not make any changes to the pluginFile variable
(the variable, not the file) and the & tells the compiler not to make a copy of it.

Add these variables to the PluginProcessor.h private data members:

```
juce::AudioPluginFormatManager pluginFormatManager;
juce::KnownPluginList knownPluginList;
juce::OwnedArray<juce::PluginDescription> pluginDescriptions;
int vstFormatInd;
std::unique_ptr<juce::AudioPluginInstance> pluginInstance;
```

1. AudioPluginFormatManager: provides functionality to load different
   plugin formats

2. KnownPluginList: not so much a list as a list manager. Can load and
   save plugin descriptions

3. PluginDescription array: the KnownPluginList manages this list of
   known plugins

4. AudioPluginInstance: this will point to the plugin once we load it

The class names for the JUCE API are confusing here since KnownPlugin
List is not a list but a manager of a list. The actual list is the OwnedArray of
PluginDescriptions.

In the constructor, initialise the manager and locate the VST3 plugin in its
list of known formats. We will need to use that VST3 format index later, and this
is a reasonable time to locate it.

```
1 pluginFormatManager.addDefaultFormats();
2 int currInd{0}, vstInd{0};
3 for (const juce::AudioPluginFormat* f : pluginFormatManager.
     getFormats()){
4    if (f->getName() == "VST3"){
5        vstFormatInd = currInd;
6        break;
7    }
8    currInd++;
9 }
```

Now everything is set for the loadPlugin function. Here is a minimal implementation of loading a plugin:

```
1  // do not call processBlock when loading a plugin
2  suspendProcessing(true);
3
4  // remove any previously read descs
5  // so we can just use index 0 to find our desired on layer
6  pluginDescriptions.clear();
7  bool added = knownPluginList.scanAndAddFile(
8                        pluginFile.getFullPathName(),
9                        true,
10                       pluginDescriptions,
11                       *pluginFormatManager.getFormat(vstFormatInd));
12
13 juce::String errorMsg{""};
14 pluginInstance = pluginFormatManager.createPluginInstance(
15     *pluginDescriptions[0], // 0 since we emptied the list at the top
16     getSampleRate(), getBlockSize(), errorMsg);
17
18 // get the plugin ready to play
19 pluginInstance->enableAllBuses();
20 pluginInstance->prepareToPlay(getSampleRate(), getBlockSize());
21 // re-enable processBlock
22 suspendProcessing(false);
```

Some comments on that code:

1. suspendProcessing prevents any audio processing until the plugin is ready

2. scanAndAddFile checks if the sent file can be loaded as a plugin

3. added is true if we can load this plugin

4. createPluginInstance actually loads a plugin into memory

5. createPluginInstance writes its result into errorMsg

6. enableAllBuses makes sure the plugin is properly initialised

At this point, you should add some checks to the loadPlugin function; for example, if the variable 'added' ends up as false, you cannot load the plugin. If the variable 'errorMsg' is empty after you call createPluginInstance, the plugin should be ready to use.

To test out loadPlugin, locate a plugin on your system. There are default locations for plugins and these vary depending on your operating system. You can query the default location using the JUCE VST3PluginFormat's getDefaultLocationsToSearch function[1]. Find a VST3 plugin using your operating system's file browser or terminal and get the full file path. Then call loadPlugin in AudioProcessor.cpp's prepareToPlay function, as that is called when the audio system is ready to go. If you call it somewhere else, e.g. the constructor, you might not know the sample rate or the block size yet, and you need those to create the plugin. Your plugin should load, and the appropriate messages you coded earlier should appear.

16.3   Use the plugin in processBlock

Now you can load a plugin; it is time to use it to generate audio in response to MIDI input. You will need a VSTi plugin, in other words, a synthesiser, as effects units do not generate audio without an input (generally speaking). I recommend you test the system using one of the plugins you have made, e.g. the FM superknob one. I have a few different test plugins, some generating a continuous tone.

Verify you can load a synthesiser plugin and then put the following in process-Block (the only code in processBlock):

```
1  // process any MIDI captured from the on-screen
2  // piano keyboard widget
3  if (midiToProcess.getNumEvents() > 0){
4      midiMessages.addEvents(midiToProcess, midiToProcess.
       getFirstEventTime(), midiToProcess.getLastEventTime()+1, 0);
5      midiToProcess.clear();
6  }
7  // call processBlock on the hosted plugin
8  if (pluginInstance){// we have a plugin loaded
9      pluginInstance->processBlock(buffer, midiMessages);
10 }
```

This code is a good start – you can now host a plugin and hear its output. Code example 39.3.6 in the repo guide provides a fully working example.

---

[1]https://docs.juce.com/develop/classVST3PluginFormat.html

16.4 Progress check

At this point you should have a plugin that can load and host another plugin. You should be able to call processBlock on the hosted plugin and it should generate an audio signal if you pass it MIDI note data.

17

Placing plugins in an AudioProcessGraph structure

This chapter describes the process of integrating plugins into an AudioProcess-Graph structure. AudioProcessorGraph is a built-in class in the JUCE library which allows the creation of a graph with multiple plugins connected together, managing MIDI and other complexities. With this graph, it is possible to host multiple plugins simultaneously. In fact, it is the underlying technology for the AudioPluginHost plugin test environment. The chapter goes through all the steps involved in converting the existing plugin controller to work with the AudioPro-cessGraph. These steps are adding the required fields for the graph and creating MIDI and audio input/output nodes, initialising the graph, wrapping the plugin in a graph node and using the graph in the processBlock function.

17.1   What is an AudioProcessorGraph?

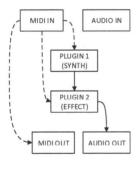

FIGURE 17.1
Wiring plugins together with a processor graph.

Before we get started, I would like to say that there are some quite large chunks of source code in this section. They are there for completeness so you can see the full implementation of graph-based audio processing in JUCE. But I would not expect you to type in all of this code from the book. You can refer to example 39.3.7 in the repo guide which has a fully working implementation of the plugin graph.

AudioProcessorGraph is a class in the JUCE API that allows you to treat audio and MIDI processing units like nodes in a graph. If you have used the AudioPluginHost plugin testing environment, you have already seen a fully developed example of what is possible with the AudioProcessorGraph. If you have not seen it, I recommend you check it out now. You can find AudioPluginHost in the extras folder of the JUCE distribution. Build it and run it. Figure 17.1 presents a simple example of a graph you can create with the AudioProcessorGraph. In that example are MIDI in and out nodes and audio in and out nodes. Two plugin nodes are embedded in the graph: a synthesizer and an effects unit.

Both plugins receive MIDI from the MIDI-in node allowing them to be controlled. The MIDI-in is also wired to the MIDI-out – a kind of MIDI-thru for those familiar with outboard synthesizers. The audio-in is not wired to anything. Audio from the synthesizer node passes to the effects unit node and then to the audio-out.

17.2   Add fields for AudioProcessorGraph

The first step is to add some new variables to the PluginProcessor to allow for the AudioProcessorGraph. You will notice that in the list of variables below there are AudioPluginInstance variables and Node :: Ptr variables. Conceptually, the Nodes are wrappers around the AudioPluginInstances. You create the AudioPluginInstance variables, then pass them to the Nodes which manage them, There are a few programming techniques going on here to make things work in a memory leak-resistant manner. Let's first take a look at the variables, which you should add to the private section of the PluginProcessor.h file:

```
1  //The graph
2  std::unique_ptr<juce::AudioProcessorGraph> audioProcGraph;
3
4  // graph components: processors
5  std::unique_ptr<juce::AudioPluginInstance>  inputProc;
6  std::unique_ptr<juce::AudioPluginInstance>  outputProc;
7  std::unique_ptr<juce::AudioPluginInstance>  midiInputProc;
8  std::unique_ptr<juce::AudioPluginInstance>  midiOutputProc;
9  // graph components: nodes that wrap the processors
10 juce::AudioProcessorGraph::Node::Ptr  pluginNode;
11 juce::AudioProcessorGraph::Node::Ptr  inputNode;
12 juce::AudioProcessorGraph::Node::Ptr  outputNode;
13 juce::AudioProcessorGraph::Node::Ptr  midiInNode;
14 juce::AudioProcessorGraph::Node::Ptr  midiOutNode;
```

The AudioPluginInstance variables are unique_ptrs and the graph nodes are Node::Ptrs. More on that shortly.

17.3 Initialise the graph

To initialise the graph, run the following in the constructor for PluginProcessor.h:

```
1  //Create basic processors for the graph
2  inputProc =
3      std::make_unique
4      <juce::AudioProcessorGraph::AudioGraphIOProcessor>
5      (juce::AudioProcessorGraph::AudioGraphIOProcessor::audioInputNode
       );
6
7  outputProc =
8      std::make_unique
9      <juce::AudioProcessorGraph::AudioGraphIOProcessor>
10     (juce::AudioProcessorGraph::AudioGraphIOProcessor::
       audioOutputNode);
11
12 midiInputProc =
13     std::make_unique
14     <juce::AudioProcessorGraph::AudioGraphIOProcessor>
15     (juce::AudioProcessorGraph::AudioGraphIOProcessor::midiInputNode)
       ;
16
17 midiOutputProc =
18     std::make_unique
19     <juce::AudioProcessorGraph::AudioGraphIOProcessor>
20     (juce::AudioProcessorGraph::AudioGraphIOProcessor::midiOutputNode
       );
21
22 audioProcGraph->enableAllBuses();
23 inputProc->enableAllBuses();
24 outputProc->enableAllBuses();
25
26 // convert the io processors into nodes
27 inputNode = audioProcGraph->addNode(std::move(inputProc));
28 outputNode = audioProcGraph->addNode(std::move(outputProc));
29 midiInNode = audioProcGraph->addNode(std::move(midiInputProc));
30 midiOutNode = audioProcGraph->addNode(std::move(midiOutputProc));
```

The code created a set of unique_ptr types, then called std::move to hand over the unique_ptrs to the nodes. We are getting into some gnarly C++ smart pointer territory here. I mentioned earlier that smart pointers manage dynamically allocated memory to ensure it is returned to the operating system when it is no longer needed. Unique_ptrs do this by only allowing one part (or one scope) of the program to access the unique_ptr at any time. Here, only the class we defined the variables in can access them. They cannot be passed around. If you want another part of the program to access a unique_ptr, you can pass it over with the move function. That is happening here: we create the unique_ptrs, then hand them over to the node objects, which take over managing them. After calling move, this part of the program can no longer access the unique_ptr variables.

Read the code carefully – can you spot any variables we need to initialise? Did you spot that audioProcGraph still needs to be initialised? Add this to the initialiser list on the constructor:

```
1    // e.g. just before #endif
2  , audioProcGraph{new juce::AudioProcessorGraph()}
```

To complete the discussion of unique_ptr, can you see the two ways we have instantiated the data stored in unique_ptr? First, we instantiated the AudioPluginInstance variables, such as midiOutputProc, by calling std::make_unique. Then, we instantiated the AudioProcessorGraph using an initialiser list. In other words, we called the constructor of AudioProcessorGraph's unique_ptr and passed it a pointer to an AudioProcessorGraph. Sheesh! The following short program illustrates these two ways of instantiating unique_ptrs, in case you re-encounter them:

```
1  #include <memory>
2
3  class Test {
4      public:
5      Test(){}
6  };
7
8  int main(){
9      // method 1: instantiate by passing a 'real pointer'
10     // to the constructor
11     std::unique_ptr<Test> myTest{new Test()};
12     // method 2: instantiate by calling make_unique:
13     std::unique_ptr<Test> myTest2;
14     myTest2 = std::make_unique<Test>();
15 }
```

17.4 Add the plugin to the graph

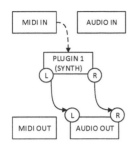

FIGURE 17.2
The graph you will
create.

To get the plugin running properly inside the graph,
you must adjust the loadPlugin function slightly. Just
remove these two lines from loadPlugin:

```
1 pluginInstance->enableAllBuses();
2 pluginInstance->prepareToPlay(
3     getSampleRate(),
4     getBlockSize()
5 );
```

Now add a function with the following signature to
the private section of PluginProcessor.h. This function
will wrap the code that does the work of wiring the
plugin into the graph as shown in figure 17.2. You can
see in the figure that the MIDI input is wired to the
plugin and the plugin is wired to the audio output. Here
is the signature:

```
1 void addPluginToGraph();
```

And an implementation in PluginProcessor.cpp, firstly,

```
1 void PluginHostProcessor::addPluginToGraph()
2 {
3     if (pluginNode) {
4         audioProcGraph->removeNode(pluginNode);
5     }
6     pluginInstance->enableAllBuses();
7
8     pluginNode = audioProcGraph->addNode (std::move (pluginInstance)
    );
9
10 }
```

Note how I used the pointer syntax to call the addNode function: instead of
audioProcGraph.addNode it is audioProcGraph− >addNode. So far so good −
we have a node sitting in the graph. Now connect the node to the appropriate
input and output nodes by adding this code to the end of the addPluginToGraph
function:

```
1 // connect the node to the output??
2 audioProcGraph->addConnection({
3     {pluginNode->nodeID, 0},
4     {outputNode->nodeID, 0}});
5 //This will crash if it's mono ... check if you want!
6 audioProcGraph->addConnection({
7     {pluginNode->nodeID, 1},
```

```
 8      {outputNode->nodeID, 1}});
 9  //Hook up midi in
10  audioProcGraph->addConnection({
11      {midiInNode->nodeID,
12       juce::AudioProcessorGraph::midiChannelIndex},
13      {pluginNode->nodeID,
14       juce::AudioProcessorGraph::midiChannelIndex}});
15
16  pluginNode->getProcessor()->prepareToPlay(getSampleRate(),
        getBlockSize());
```

Then call it from loadPlugin, before you 'unsuspend' the audio:

```
 1  ...
 2  pluginInstance = pluginFormatManager.createPluginInstance(
 3      *pluginDescriptions[0],
 4      getSampleRate(), getBlockSize(), errorMsg);
 5  if (errorMsg == "") {  DBG("Plugin probably loaded");}
 6  else {DBG("Plugin failed to load " << errorMsg); return;}
 7
 8  // then:
 9  addPluginToGraph();
10  suspendProcessing(false);
11  // end of loadPlugin
```

17.5   Call the graph in processBlock

The final step in moving to the graph model is to call processBlock on the graph in your host's processBlock function. This is very similar to how you did it with the pluginInstance except instead of calling processBlock on pluginInstance you call processBlock on audioProcGraph. Edit the processBlock function in Audio-Processor.cpp:

```
 1  ... midi stuff.
 2
 3  if (pluginNode){
 4      audioProcGraph->processBlock(buffer, midiMessages);
 5  }
```

You should now be able to load the test plugin and hear it via the graph.

17.6 Load any plugin

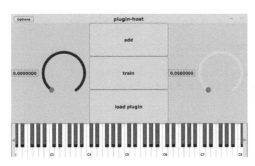

FIGURE 17.3

User interface for the basic host with the load plugin button added.

Now you will add a new button to the plugin interface that lets you load any plugin. Add a button to PluginEditor.h in the private section and follow the usual steps to make it appear: addAndMakeVisible, addListener, and update the resized function. You should end up with a user interface like figure 17.3. The new button should call the buttonClicked callback on the PluginEditor. But how to connect the button to the loadPlugin function? We need to trigger a file browsing dialogue, then collect the resulting filename and pass it to loadPlugin. The following code will show a file dialogue. Put it in the buttonClicked callback when the button is the load plugin button:

```
1 auto fileChooserFlags = juce::FileBrowserComponent::openMode |
2                         juce::FileBrowserComponent::canSelectFiles;
3
4 fChooser.launchAsync(fileChooserFlags,
5                 [this](const juce::FileChooser& chooser){
6     DBG("you chose a plugin file " << chooser.getResult().
        getFullPathName());
7     this->audioProcessor.loadPlugin(chooser.getResult());
8 });
```

To make it work, you need to add a private variable to the PluginEditor.h file:

```
1 juce::FileChooser fChooser{"Select a plugin."};
```

The code uses an anonymous function as an asynchronous callback. If you are familiar with Javascript, this pattern is familiar to you. The idea is that the GUI thread is not interrupted by the user browsing for a file, as the file browser happens in a separate thread. Once the user has selected a file, the code prints out the file path and then attempts to load it using the AudioProcessor's loadPlugin function. The fileChooseFlags specify that the user can select files. The problem is that plugins do not present themselves as files on all platforms. On some platforms they present themselves as directories.

If you want your application to be cross-platform, you need to specify different flags for different platforms. MacOS and Linux plugins are directories. Windows plugins are files. So you need to set the flags using some JUCE macros to check which platform you are on.

```
1  # ifdef JUCE_LINUX
2  auto fileChooserFlags = juce::FileBrowserComponent::openMode |
3                  juce::FileBrowserComponent::canSelectDirectories;
4
5  # endif
6  # ifdef JUCE_MAC
7  auto fileChooserFlags = juce::FileBrowserComponent::openMode |
8                  juce::FileBrowserComponent::canSelectFiles;
9  # endif
10 # ifdef JUCE_WINDOWS
11 auto fileChooserFlags = juce::FileBrowserComponent::openMode |
12                  juce::FileBrowserComponent::canSelectFiles;
13 # endif
```

17.7 Progress check

At this point you should be able to load any plugin and play it using the piano control on the host application. Code example 39.3.7 in the repo guide provides a fully working example of the above.

18

Show a plugin's user interface

The final functional change we will make to our plugin host before we switch back to working on the machine learning side is to display the user interface for the plugin. This chapter explains how to display the user interface for any plugin. One of the features of the various plugin APIs (VST3, AudioUnit, etc.) is that they allow the plugin to have a custom user interface. This is the interface you see when you load the plugin into your DAW. As you should know by now, JUCE allows you to write a custom GUI for your plugins using the JUCE GUI widgets. Commercial and other plugins have their own ways of implementing the GUI, including animated graphics, 3D, etc. How can we show this user interface using JUCE? The solution presented here is to pop up a free-floating window containing the user interface. To achieve that, you are going to explore the JUCE API a little more and define a new class.

18.1 AudioProcessor and AudioProcessorEditor class hierarchies

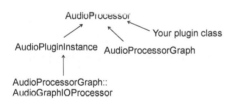

FIGURE 18.1
Class hierarchy for AudioProcessor and its descendants.

At this point, I will take a little detour to discuss the class hierarchy for AudioProcessors, AudioPluginInstance, AudioProcessorGraph and so on. This will allow you to better understand how to go about launching a plugin's user interface from a plugin embedded in an AudioProcessorGraph inside a node. Figure 18.1 shows the class hierarchy for the AudioProcessor classes. As you can see AudioPluginInstance and AudioProcessorGraph are both AudioProcessors. So is the application we are working on now, which

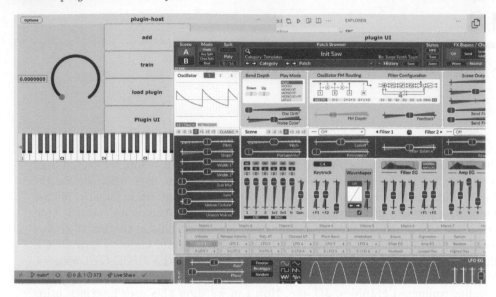

FIGURE 18.2
User interface for the host with the Surge XT plugin user interface showing in a separate window.

is also a plugin. This is why they all have processBlock, prepareToPlay and so on – they inherited them from AudioProcessor.

Another thing that AudioProcessor classes have are various functions for retrieving the user interface. Check out your PluginProcessor.h – there is a function override called createEditor. Look at the implementation in PluginProcessor.cpp – it creates and returns an instance of the editor class defined in your PluginEditor files. In my PluginProcessor.cpp file, the createEditor function contains one line like this:

```
return new PluginHostEditor (*this);
```

If I switch to my PluginEditor.h, I can see that it defines a PluginHostEditor class, and it takes an AudioProcessor to its constructor:

```
PluginHostEditor (PluginHostProcessor&);
```

So whenever a plugin host wants to see the user interface of my plugin, somehow it calls createEditor and createEditor calls the constructor of the user interface class and returns it.

What is the base class of the editor? In my PluginEditor.h file, the class inheritance is defined as follows:

```
class PluginHostEditor :
```

```
2     public juce::AudioProcessorEditor,
3     public juce::Button::Listener,
4     public juce::Slider::Listener,
5     private juce::MidiKeyboardState::listener
```

So in fact it has multiple base classes. The one we are interested in here is AudioProcessorEditor. If you look at the class hierarchy for AudioProcessorEditor, you will find it inherits from Component. A Component, just like any other Component in JUCE, such as a TextButton or a Slider. You could display it directly on your user interface. Similarly, if you called createEditor on any external plugin you have loaded, you will receive an AudioProcessorEditor object and it too would be a Component. So if you want to display the user interface for a plugin, you could just treat it as a Component and place it directly into your interface window.

But embedding the plugin UI in your main window is not a good idea: the lifetime of plugin UI components is a little odd, so you do not want to rely on it as a part of your main UI. The plugin UI can have a completely different aesthetic to your UI, so it does not really make sense to simply place it in your UI. And the plugin's user interface will have its own preferred size. In the following section, you will find out how to create a separate window for the plugin UI.

18.2   Accessing the plugin's user interface

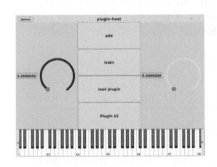

FIGURE 18.3
User interface for the host with a show UI button.

To create a new window for the plugin UI you will need a button on your main UI to open the plugin window and then you will need a new class to represent the window which you can instantiate and render the plugin UI into. Start by adding a new button to the main UI. The button will eventually trigger the opening of the loaded plugin's UI. As usual, you will need to call addAndMakeVisible to add it to the UI, call addListener to assign the main UI as a listener to the button then adjust the size in the resized function. My user interface ended up looking like figure 18.3.

Once you have the button displaying and you are receiving events when it is clicked, you need to figure out how to access the user interface for the plugin. Remember that you wrapped the plugin in a graph node when you implemented the AudioProcessorGraph. So you need to somehow extract the plugin 'AudioProcessor' from the graph. Once you have access an AudioProcessor representing the plugin, you can call its createEditor function to gain access to the plugin UI.

So in PluginProcessor.h, add a new public function which will ultimately return an AudioProcessorEditor object representing the plugin's UI:

```
juce::AudioProcessorEditor* getHostedPluginEditor();
```

Then the implementation in PluginProcessor.cpp:

```
juce::AudioProcessorEditor* PluginHostProcessor::
    getHostedPluginEditor()
{
    if (pluginNode){
        return pluginNode->getProcessor()->createEditorIfNeeded();
    }
    else{
        return nullptr;
    }
}
```

On line 3 the code checks if a pluginNode is available. This will only be the case if you have already loaded a plugin via the loadPlugin button. If the plugin-Node is available, it calls getProcessor on it, which will return a pointer to the AudioProcessor inside the node, or in other words, the plugin. This is odd since earlier you wrapped the plugin in a smart unique_ptr and now it is being converted back to a regular bare pointer. I think at this point you just need to be a good citizen and not call delete on that pointer. So how do you get the UI from an AudioProcessor? As mentioned earlier, you call createEditor, or better, createEditorIfNeeded which will only call createEditor if necessary. Right so you have access to the AudioProcessorEditor for your hosted plugin. Now you need to get that displayed in a window.

18.3   The PluginWindow class

Now for the PluginWindow class itself. Add two new empty files to your project: PluginWindow.h and PluginWindow.cpp. You will need to add at least the cpp file to your CMakeLists.txt file, as you did earlier for the NeuralNetwork class. That involves editing the target_sources command:

```
    target_sources(fm-torchknob
        PRIVATE
        src/PluginEditor.cpp
        src/PluginProcessor.cpp
        src/NeuralNetwork.cpp
```

```
6    stc/PluginWindow.cpp)
7
```

Now put some code into the PluginWindow.h file:

```
1 class PluginWindow  : public juce::DocumentWindow
2 {
3 public:
4     PluginWindow (juce::AudioProcessorEditor* editor);
5     ~PluginWindow() override;
6     void closeButtonPressed() override;
7 // standard macro call used to make components play nicely
8 JUCE_DECLARE_NON_COPYABLE_WITH_LEAK_DETECTOR (PluginWindow)
9 };
```

Here are the definitions of all those functions for PluginWindow.cpp:

```
 1 #include "PluginWindow.h"
 2
 3 PluginWindow::PluginWindow (juce::AudioProcessorEditor* editor)
 4     :
 5     DocumentWindow ("plugin UI",
 6         juce::LookAndFeel::getDefaultLookAndFeel().
 7         findColour (juce::ResizableWindow::backgroundColourId),
 8         juce::DocumentWindow::minimiseButton |
 9         juce::DocumentWindow::closeButton)
10 {
11     setSize (400, 300);
12
13     if (editor){
14         setContentOwned (editor, true);
15         setResizable (editor->isResizable(), false);
16     }
17     setTopLeftPosition(100, 100);
18     setVisible (true);
19 }
20
21 PluginWindow::~PluginWindow()
22 {
23     clearContentComponent();
24 }
25
26 void PluginWindow::closeButtonPressed()
27 {
28 }
```

Now you should be able to use the PluginWindow class to open a window showing the plugin user interface. In PluginEditor.h, add this to the private fields:

```
1 std::unique_ptr<PluginWindow> pluginWindow;
```

Then in PluginEditor.cpp, implement the code to create the window when the user presses the show UI button:

```
1  if (btn == &showPluginUIButton){
2      if (!audioProcessor.getHostedPluginEditor()) {
3          DBG("showPluginUIButton no plugin");
4          return;
5      }
6      if (pluginWindow){
7          DBG("showPluginUIButton window already open");
8          return;
9      }
10     DBG("showPluginUIButton creating new gui");
11     pluginWindow =
12         std::make_unique<PluginWindow>
13             (audioProcessor.getHostedPluginEditor());
14 }
```

Note that the code deals with various possible scenarios that can occur: there is no plugin loaded yet, the plugin is loaded but the window already exists and the plugin is loaded and the window does not exist.

Try it out! You should see something like figure 18.2, which shows the Surge XT user interface rendered into a PluginWindow alongside the host user interface. Of course, you will see the user interface for your plugin of choice.

18.4   Closing the plugin user interface window

You may have noticed that closing the plugin user interface window is impossible. To close the window, you need to implement the closeButtonPressed function. The following instructions show you how to do this in a 'JUCE' way by defining a custom PluginWindowListener abstract class you can inherit and implement to receive the close event in the PluginEditor. First, add the following abstract class to the top of PluginWindow.h:

```
1  class PluginWindowListener{
2  public:
3    PluginWindowListener(){}
4    ~PluginWindowListener(){}
5    virtual void pluginCloseButtonClicked() = 0;
6  };
```

Can you work out what makes it an abstract class instead of a normal one? That should be an easy question if you are a seasoned C++ programmer. If not, the answer is that PluginWindowListener is an abstract class because plugin-CloseButtonClicked has its implementation set to '0'. You cannot instantiate an abstract class, so you cannot directly use it. This forces any class that inherits from PluginWindowListener to either be abstract itself or to provide an implementation for that function. JUCE makes extensive use of abstract classes to force you to implement things. For example, the Button::Listener that you have been inheriting from has its buttonClicked function set to '0', forcing you to implement it in your listener.

Back to the task at hand, add a private variable to the PluginWindow class:

```
PluginWindowListener*  listener;
```

And a function to assign a listener (you can put this into PluginWindow.h):

```
void addPluginWindowListener(
    PluginWindowListener*  _listener)
    {
        listener = _listener;
    }
```

Then notify the listener when the PluginWindow's close button is pressed in PluginWindow.cpp:

```
void PluginWindow::closeButtonPressed()
{
    if (listener){
        listener->pluginCloseButtonClicked();
    }
}
```

Now you need to create a Listener that can listen to the close event and do the things necessary to remove the window. Inherit from PluginWindowListener in the PluginEditor:

```
class PluginHostEditor  :    public juce::AudioProcessorEditor,
                             public PluginWindowListener,

... in the public section:
  void pluginCloseButtonClicked() override;
...
```

Add the implementation in PluginEditor.cpp:

```
void PluginHostEditor::pluginCloseButtonClicked()
{
    // trigger delete of unique_ptr-owned object
    pluginWindow.reset();
}
```

The last step is to register the PluginHostEditor as a listener. In PluginEditor.cpp's buttonClicked function, after you create the window, register as a listener:

```
pluginWindow = std::make_unique<PluginWindow>(audioProcessor.
    getHostedPluginEditor());
pluginWindow->addPluginWindowListener(this);
```

18.5   Progress check

At this point, you should be able to load a plugin then click a button on your main UI that pops up a new window displaying the plugin's UI. You should also be able to close the plugin UI window, which was perhaps the hardest part!

19

From plugin host to meta-controller

This chapter completes the implementation of the plugin meta-controller by explaining how to integrate the plugin host with the neural network code from the torchknob. The first section discusses querying plugin parameters and filtering out un-needed parameters. The following section covers creating a neural network for the plugin with the correct number of inputs and outputs. The next section discusses creating training data by setting the training knob position and selecting plugin presets. The section following that explains how to control the plugin using the neural network. The chapter concludes by mentioning possible extensions for the meta-controller, such as training dataset population, custom UI components, selective parameter control, multiple synthesizer hosting, and improved usability with state loading/saving and input features from other sources.

19.1 Querying plugin parameters

FIGURE 19.1
User interface for the Dexed DX-7 emulator. It has 155 parameters.

Think back to the design of the torchknob FM synthesis control system. The knob controlled two parameters at the same time. Now you will scale that control to all parameters on the plugin synthesizer. But how do you know what the parameters are on a given plugin? I have been abusing plugins with custom hosts for some time. For example, I worked with Martin Roth to develop an automatic VST plugin programmer in 2008[47]. Back then, you could ask the plugin for a list of parameters, and you would receive an array of strings and metadata for the parameters. Then you could set parameters by passing float values in the range 0-1. This is more or less how it works in JUCE now. But things are a made a little more complex because the plugin wrapper in JUCE, the AudioProcessor class[1], is plugin format agnostic. In other words, it uses the same set of parameter control functions for multiple plugin formats. Some formats have slightly more complex ways of arranging parameters, such as groups of related parameters. That makes the plugin querying code a little more complicated.

The following code prints a list of plugin parameters and their values. Add it to PluginProcessor.h and cpp as a private function:

```
void PluginHostProcessor::printPluginParameters()
{
    if (pluginNode){

        juce::Array<juce::AudioProcessorParameter *> params =
        pluginNode->getProcessor()->getParameters();

        for (const juce::AudioProcessorParameter* p : params ){
            DBG("p: " << p->getName(100) << " : " << p->getValue());
        }
    }
}
```

The '100' parameter for getName tells it to automatically crop the name to 100 characters. You have to pass this parameter. Add the code to your PluginProcessor and call printPluginParameters after loading a plugin. Here is an extract of the output I see when I load the Dexed plugin:

```
p: Cutoff : 1
p: Resonance : 0
p: Output : 1
p: MASTER TUNE ADJ : 0.5
p: ALGORITHM : 1
p: FEEDBACK : 1
p: OSC KEY SYNC : 1
p: LFO SPEED : 0.353535
p: LFO DELAY : 0
p: LFO PM DEPTH : 0
p: LFO AM DEPTH : 0
p: LFO KEY SYNC : 1
p: LFO WAVE : 0
p: TRANSPOSE : 0.5
p: P MODE SENS. : 0.428571
p: PITCH EG RATE 1 : 1
p: PITCH EG RATE 2 : 1
```

At the end of the output, I see many parameters I did not expect, for example:

```
p: MIDI CC 15|125 : 0
p: MIDI CC 15|126 : 0
p: MIDI CC 15|127 : 0
p: MIDI CC 15|128 : 0
```

---

[1] https://docs.juce.com/master/classAudioProcessor.html

If you develop a plugin using JUCE, JUCE adds extra parameters by default. I believe the purpose of these parameters is to expose the plugin to MIDI CC control somehow. But for now, we need to ignore them as they are not the core parameters we wish to manipulate. You can simply add a filter to the print function to remove these special parameters:

```
for (const juce::AudioProcessorParameter* p : params ){
    if (! p->getName(100).contains("MIDI CC ")){
        DBG("p: " << p->getName(100) << " : " << p->getValue());
    }
}
```

19.2   Create a neural network for a plugin

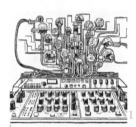

Now you have the correct set of parameters printed out, you are ready to create a neural network that can control them. The network will have one input for the torchknob and one output for each parameter. First, add a public function similar to the printPluginParameters function to PluginProcessor.h. Its purpose is to count the number of valid parameters. Here is a signature:

```
int countValidParameters();
```

FIGURE 19.2
Time for a more complex neural network.

You need to write the implementation, using the same filtering as printPluginParameters.

The NeuralNetwork object is stored in the PluginEditor. This makes sense as the PluginEditor receives the knob movements and then passes them to the NeuralNetwork. So after the PluginEditor calls loadPlugin, it should query the number of parameters and configure the network accordingly. So in PluginEditor.cpp::buttonClicked, after you call loadPlugin, set up the neural network:

```
fChooser.launchAsync(fileChooserFlags ,
    [this](const juce::FileChooser& chooser){
        this->audioProcessor.loadPlugin(chooser.getResult());
        this->nn = NeuralNetwork(
            1,
            this->audioProcessor.countValidParameters()
        );
});
```

That creates a neural network with one input and as many outputs as the plugin has parameters. You will need a variable in the PluginEditor.h private variables called nn and of type NeuralNetwork. Verify that your code builds and runs. You cannot train the network yet, as you must also create training data from the plugin.

19.3 Create training data from plugin parameters

Remember that the training data consists of inputs which are values read from the torchknob slider (or other GUI elements if you worked on the challenges from the torchknob section). The outputs consist of sets of parameter settings for the synthesizer. To generate input-output pairs for the training data, you must read the torchknob slider value and the current parameter settings. Write a public function in PluginProcessor.h/cpp that returns a vector containing the parameter values for all valid parameters on the plugin. Here is a signature for the function:

```
std::vector<float> getHostedPluginParamValues();
```

Write your own implementation, referring to the parameter querying code you can find in the printPluginParameters function above.

When the user clicks the add button, call addTrainingData on the NeuralNetwork. Here is the signature for the addTrainingData function:

```
    void addTrainingData(
        std::vector<float>inputs,
        std::vector<float>outputs);
```

You should call it in the buttonClicked function in PluginEditor.cpp. The call should look something like this:

```
nn.addTrainingData(
        {(float) superKnobTrain.getValue()},
        this->audioProcessor.getHostedPluginParamValues());
```

You should now be able to adjust the torchknob, adjust the parameters on the plugin and add training data. The effects of this will all be invisible until you activate the other dial on the right-hand side of the interface.

19.4 Control the plugin with the neural network

The final step is to activate the control knob on the right-hand side of the interface. The control knob should send a value into the neural network, then take the outputs and send them to the plugin's parameters. Create a public function in PluginProcessor.h:

```
void setHostedPluginParamValues(std::vector<float> values);
```

This is the opposite of getHostedPluginParamValues. The implementation works like this:

```
void PluginHostProcessor::setHostedPluginParamValues(std::vector<
    float> values)
{
  int pind{0};
  if (pluginNode){
  juce::Array<juce::AudioProcessorParameter *> params = pluginNode->
    getProcessor()->getParameters();
    for (const juce::AudioProcessorParameter* p : params ){

      if (! p->getName(100).contains("MIDI CC ")){
        pluginNode->getProcessor()->setParameter(pind, values[pind]);
        pind++;
      }
    }
  }
}
```

That is all for now. Experiment by moving the training knob, selecting a preset on the plugin, hitting add, and repeating it a few times. Then see how well the plugin morphs between the presets. You will notice that setParameter is considered to be deprecated in JUCE. You are supposed to use AudioProcessorParameter and AudioProcessorParameterGroup, but this works for now and is simpler.

19.5   Extending the meta-controller

There are many possible extensions you can consider for the meta-controller:

1. Automatically populate a training dataset by iterating over the presets on the plugin, loading them and assigning them to different positions on the knob

2. Custom UI components to provide control input. Instead of one knob, you could try to implement an X,Y pad

3. Use MIDI CC data as the control input

4. Allow the user to select which parameters they want to control

5. Host and control multiple synthesizers

6. Improved usability: loading and saving state and possibly the model

7. Take input control values from another, higher dimensional source, e.g. extract features from an audio feed's spectrum or even a video feed.

19.6   The end of this part of the book

Excellent work! You have reached the end of this part of the book, wherein you have been working on the meta-controller. I hope you have found some unexpected places in the high-dimensional latent space of synthesizer parameters. You can, of course, use the components developed here in any way you see fit. You might find the plugin hosting a plugin concept sends you off in another direction entirely. Or you might find the many-many trainable mapping part can apply to another musical or even non-musical problem.

Part III

The autonomous music improviser

20

Background: all about sequencers

In this part of the book, you will build an autonomous music improviser. The music improviser is a plugin that receives MIDI data, models the data and generates MIDI data in the style of the previously received input data. You can think of interacting with the improviser as having a musical conversation with yourself. Before we get into the details of building the improviser, in this chapter, I will provide some background information on the field of algorithmic composition and its related offshoot, musical agents. You will discover a vast treasure trove from decades of research into analysing and generating music. I will then explain the algorithm you will use to implement the improviser: the variable order Markov model. I will compare the model to others, such as deep learning systems, and explain why I have chosen this technique.

20.1 Beyond the sequencer

FIGURE 20.1
How far are modern sequencers from steam-powered pianos?

I would like to start by placing the improviser in a kind of 'taxonomy' of music technology. In present-day music production, it is generally accepted that the primary forms of technology are covered by the following list: traditional instruments, digital audio workstations (DAWs), plugins, synthesisers, sequencers, samplers, mixing tools, recording hardware and MIDI and other controllers[36, 1]. These categories are not mutually exclusive – some technology crosses into several categories, and the technology might be software or hardware. I would place the improviser in the 'sequencer' category, though it will also be a plugin. But, as per the title of this section, it goes beyond the sequencer – it is a kind of smart sequencer which can generate new sequences based on the sequences you feed it.

FIGURE 20.2
My own experience interacting with AI improvisers. Left panel: playing with Alex McLean in Canute, with an AI improviser adding even more percussion. Right panel: livecoding an AI improviser in a performance with musician Finn Peters.

To contextualise the improviser more fully, we need to look beyond commercial music technology towards research into new music technology. Algorithmic composition is a well-established area of research in music technology, and it involves the generation of musical sequences via some specified procedure. Stories of algorithmic composition's rich history commonly start with Mozart's dice game, wherein the composer writes several fragments of music and then assigns each fragment a number from 1 to 6. A series of dice rolls select the order in which the fragments appear in the resulting composition.

Fast-forwarding to the 20th century, the first composition generated algorithmically using a computer is generally reported as the 'Illiac Suite' by Hiller and Isaacson in 1957[16]. The Illiac Suite utilised several algorithmic composition techniques, including sampling musical notes from probability distributions and Markov models, which select notes based on previous notes. You will learn much more about Markov models as you work through the following few chapters. Other composers who pioneered the use of computers and saw the potential of algorithmic processes such as probability-based methods include Xenakis and Koenig. You should seek out Koenig's works, such as Funktion Blau – there are some fascinating textures and patterns to be heard.

The development of algorithmic composition from the Illiac Suite until now follows along with the phases of development seen in artificial intelligence and machine learning. After those early experiments in the 50s and 60s, the 1970s saw the development of formalisms such as musical grammars, which allowed the definition of rules to identify and generate stylistically correct music[37, 19]. The 1980s saw the development of more advanced Markov modelling techniques[8], building on work from speech recognition research. The work on Markov modelling continues until the present day, though many researchers are switching to deep neural network methods. Pachet et al. have done some of the most sophisticated work with Markov models, using them to carry out musical style transfer in the Flow Machines project in the 2010s[13].

The 1990s saw the use of dynamical systems techniques, chaos and cellular automata, e.g.[27]. The 1990s also saw algorithmic composition researchers using evolutionary computation, the emerging trend in computational problem-solving at the time. For example, Biles' GenJam jazz improvisation generator[3]. As an aside, I began experimenting with evolutionary techniques applied to sound synthesiser control during my master's degree in the late 1990s. I reimplemented some of that work using the Web Audio API in the mid 2010s[1]. I then went on to carry out my doctoral research using evolutionary techniques to program synthesisers[48], eventually switching to deep neural network techniques in the late 2010s[49]. Automated sound synthesizer programming is not exactly algorithmic composition, though!

Returning to the main thread, we have reached the early 2000s, which saw the culmination of techniques such as musical grammars, genetic algorithms and Markov modelling. This is very much a potted history – I do not want to get too bogged down in a literature review here. For further reading, there are many review articles available in open access, e.g.[15] and Nierhaus's 2009 book provides plenty of detail on techniques through to the deep learning era[30]. Speaking of deep learning, many areas of applied computer science, including algorithmic composition, entered the deep learning era in the 2010s, and there has been a lot of work on algorithmic composition since then. In fact, it has been an 'endless stream' according to Ji et al. in their 2020 comprehensive review of deep music generation[21]. As well as that review paper, Briot's article series reviewing deep learning techniques for music generation is an excellent starting point for further reading[4].

In summary, there have been decades of work in algorithmic composition utilising a range of different computational techniques. Now that I have given you some idea of the depth and range of that work, I will zoom in on music improvisers which is our area of interest for the next few chapters.

20.2 Musical agents

I will start by introducing you to the idea of a 'musical agent'. In their 2019 review of research on musical agents, Tatar and Pasquier describe musical agents

---

[1] http://www.yeeking.net/evosynth

as "artificial agents that tackle musical creative tasks, partially or completely". A related area of work is 'musical meta-creation', which explores the idea that instead of creating music directly, you create systems that can generate music in some way. I think musical meta-creation is a rebranding of algorithmic composition, which was probably looking a bit crusty to new researchers coming into the field in the 2010s[2].

In the taxonomy paper I mentioned above, Tatar and Pasquier describe many aspects of musical agent design, and we will be considering several of those as we develop the improviser plugin. As a starting point, some questions to consider when designing a musical agent are: what is the musical corpus it uses to generate from? Is the corpus raw audio or symbolic (e.g. MIDI) data? What level of autonomy does the system have, and which part of the composition process does it carry out?

I can answer some of those questions for the improviser plugin you will work on through the following few chapters. The musical corpus will consist of MIDI data that you feed into the model during a performance, though it can also save and load models. The corpus is symbolic as it consists of MIDI data representing musical events instead of raw audio. The system will have a high level of autonomy, at least in how we will implement it here. It will choose when and what to play, but it will depend on its human partner to provide the data it learns from. Regarding the composition process, the improviser is a kind of composition aid that helps generate ideas. But it is also a great tool to use in live performance.

Speaking of live performance with autonomous improvisers, I will finish this pre-amble by saying that I have worked with AI improvisation systems for many years. In figure 20.2, you can see two examples: firstly, I played a drumkit along with livecoding pioneer Alex McLean and an AI percussionist which learned from and responded to my drumming. In the other photo, I am actually writing an AI improviser in real-time using the SuperCollider language during a livecoding performance with Finn Peters, with whom I have collaborated on many projects, including his Butterflies and Music of the Mind albums[3]. One of the Finn Peters and AI improviser performances was recorded for Jez Nelson's BBC Radio 3 'Jazz on 3' programme, and even included in their list of sessions of the year[4].

The system we will develop in this part of the book is one of a series of systems I have designed. I mentioned the performances with Finn Peters, but other versions of this improviser have performed live on BBC national radio[5] and been used by other musicians such as pianist Carles Marigo, who created a stage show called Bach to the Future based around the technology in 2022.

---

[2]My apologies to the very talented Arne Eigenfeldt for that comment.
[3]https://www.discogs.com/artist/343757-Finn-Peters
[4]https://www.bbc.co.uk/programmes/b00pdwgn
[5]https://www.bbc.co.uk/programmes/p033s4gj

20.3   Sequence modelling

The basic idea behind the improviser we will develop is sequence modelling. Sequence modelling means trying to understand the properties of a sequence. With that understanding, you can carry out tasks like predicting the next thing in the sequence or generating sequences similar to those in a given dataset. The improviser aims to model sequences that you play into it such that it can then generate similar but interestingly different sequences in response. Sequence modelling is a huge research field in computer science. It has applications in many domains, such as financial data modelling, natural language processing and audio signal modelling. In fact, large language models are sequence modellers. GPT and similar models ultimately aim to predict the next symbol in a sequence.

There are many techniques available to describe and model sequences. For example, you might define a mathematical relationship such as 'the next number in a sequence of numbers will be double the previous number'. That would be a great model if it described your bank account. Or you might describe some rules, such as 'if you have a 1, the next number is always 3, if you have a 3, the next number is always 1'. That would yield a sequence of 1,3,1,3... Moving to the musical domain, you can come up with musically meaningful rules such as 'the sequence can only contain notes in the key of C'. To generate sequences from that rule, you can just randomly select notes from the key of C.

These kinds of clear rules are easy to understand, but the problem is that many real-world sequences are too complex or noisy to allow simple rules for prediction and generation. This is where *statistical* sequence modelling comes in. Statistical sequence modelling aims to identify patterns in sequences that include probabilistic elements. For example, 'if the last number was 1, there is a 50% chance that the next number will be 2 and a 50% chance it will be 4'. In fact, that is a first-order Markov model. First order because it only looks one step back in the sequence. You can even have a zero-order Markov model, commonly called a probability distribution. A zero-order Markov model might say: 'There is a 50% chance of a 4, a 25% chance of a 2 and a 25% chance of a 1'. Zero order means it does not look at the past – it just looks at the raw distribution of numbers observed in the sequence.

20.3.1   Deep networks as sequence modellers

Before we dive deeper into Markov models in the next section, I will say that deep neural networks that generate musical or other sequences like those reviewed in Ji et al.'s comprehensive review[21] are also fantastic statistical sequence modellers. You could use a deep neural network to build a musical improviser. In fact, deep neural networks are technically better sequence modellers than Markov models – this is why GPT and Bard are neural networks, not Markov models!

However, the problem with deep neural networks, and the reason I have yet to start using them to build improvisers, is that they need a lot of data and time to train, and they cannot take on new data in real-time. They can respond to your input in real-time, but the underlying model will not change or learn in real-time. To change the underlying model requires a training phase. If you consider what the training data actually is, you might encounter some problems. For example, do you want your improviser to improvise using a massive dataset of music written by other people? Perhaps you do, maybe you do not, but with deep networks, you do not have a lot of choice.

Markov models can generate interesting patterns from data gathered from scratch in real-time from a single performance if you want, or you can feed them larger datasets. The improviser we will build is computationally efficient and can learn on a small dataset in real-time. Musicians I have worked with clearly feel their own presence in the music generated by these kinds of models and enjoy interacting with them. I am not saying you cannot achieve this with neural networks, but you can certainly accomplish this with Markov models, so that is what we shall use.

20.4   Markov models

So, we are going to use Markov models to model musical sequences. In the previous section, I hinted at how Markov models work and mentioned that they can have different orders. Let's start with a sequence consisting of just two different musical notes. Imagine you have a sequence as follows:

$$A, A, B, A, B, A, B, B, B \qquad (20.1)$$

To create a Markov model, we start by calling each of those letters a *state*. So, at steps 1 and 2 in the sequence, the state is 'A'; at step 3, the state is 'B'. Then we talk about *state transitions*. From step 1 to step 2, the state transitions from A to A. From steps 2 to 3, the state goes from A to B. It is true that A to A is not strictly a transition, as it looks like nothing changed, but imagine you are playing musical notes. You play an A, then another A, then a B and so on. There are 8 transitions in that sequence:

$$A \to A, A \to B, B \to A, A \to B, B \to A, A \to B, B \to B, B \to B \qquad (20.2)$$

We can refer to the state after the transition as an *observation*. From those

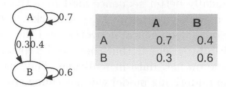

FIGURE 20.3
Visualisation of a two-state model on the left and the state transition probability table on the right.

transitions, we can compute the probabilities for an observation given the current state. So, state A is followed by either observation A or B. The observation A follows state A once. State A goes to observation B three times. Therefore, the $A \rightarrow A$ transition has a probability of 25%, and the $A \rightarrow B$ has a probability of 75%. We can compile all of that into a *state transition matrix* showing the probabilities of transition between all observed states:

$$
\begin{array}{ccc}
 & A & B \\
A & 0.25 & 0.75 \\
B & 0.5 & 0.5
\end{array}
\tag{20.3}
$$

Figure 20.3 illustrates this in the form of a graph, with the state transition matrix alongside for reference.

20.4.1 Generating from the model

The next step in this thought experiment is to generate a sequence from the model. This starts with selecting a starting state, e.g. A. Then, you sample from the distribution of possible observations, so you will either receive A or, more likely, B. The letter you choose now becomes the state. You can then use the state transition matrix to identify all possible observations following that state. The observation chosen then becomes your next state. And so on.

That's it – if that made sense to you, then you now understand first-order Markov models. If you are not sure, that's ok, we are going to get into some hands-on work in C++ in the next chapter. Since I mentioned them earlier, what would a zero-order Markov model of this sequence look like? Zero order is just a distribution of all observed states. We saw four As and five Bs. Generating from a zero-order model just means sampling from that distribution. So you have a 44.4% chance of an A and a 55.5% chance of a B each step.

Now it is your turn – write out a sequence of musical notes, then try to convert it into a state transition matrix containing probabilities of each possible transition. Then, try to generate a new sequence using your state transition matrix.

20.5   Higher order models

The next step up in complexity is to increase the order of the model beyond the first order. First order means we look at one previous step. Second order looks at two previous steps, but how does that work?

Going back to our original sequence:

$$A, A, B, A, B, A, B, B, B \tag{20.4}$$

Second-order modelling looks at the observations following states that are made from two letters:

$$A, A \rightarrow B$$
$$A, B \rightarrow A$$
$$B, A \rightarrow B$$
$$A, B \rightarrow A$$
$$B, A \rightarrow B$$
$$A, B \rightarrow B$$
$$B, B \rightarrow B$$

As a state transition matrix, you have:

	A	B
A,A	0	1
A,B	0.666	0.333
B,A	0	1
B,B	0	1

(20.5)

To generate from this second-order model, you would typically choose a two-letter state from the left column, then sample from its possible next states. The next state you choose is then pushed on to the end of the previous state, and the older state is dropped, so you retain a two-letter state. In terms of data structures, this is a FIFO stack. With the probabilities in the matrix above, generated sequences would not be particularly interesting as you would usually end up generating lists of Bs. Sequence modellers getting stuck in a repetitive state like this in a musical context is not desirable. Well, maybe for techno fans. I will explain how I work around that problem shortly.

You can operate in whatever order you like, up to the length of your sequence

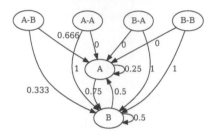

FIGURE 20.4
Visualisation of a variable order Markov model containing first and second order states.

minus one. As the order increases, the number of observations decreases. For example, if you have nine letters in your sequence like we did, you can only make one eighth-order observation. Higher order means your model knows more about the history of events before an event, so it might make more sensible musical decisions. To clarify that with a musical example, imagine you were doing a listening test where you listen to a fragment of music, then you have to guess what the next note or chord is. Do you think you would find it easier to guess the next note with a longer or shorter fragment?

However, higher order also means fewer observations, meaning the model has less variation in its output. How can you make sure the model has both the stateful, musical awareness provided by higher orders and the variation that makes for interesting sequences? In the next section, I will describe the variable order Markov model, which takes advantage of both the richness of lower-order models and the understanding of higher-order models.

20.6   Variable order models

Variable order Markov models maintain models at several different orders simultaneously. Building variable order models is similar to building fixed order models except you compute matrices for all orders and combine them into a single matrix. A combination of our first- and second-order models would look like this:

	A	B
A	0.25	0.75
B	0.5	0.5
A,A	0	1
A,B	0.666	0.333
B,A	0	1
B,B	0	1

$$(20.6)$$

As you can see, with the same sequence, we have the richness of the first-order model (more observations) with the information of the second-order model (longer memory). Figure 20.4 presents a graph-based visualisation of this model. Whilst the model looks more complex, it is just a combination of the first- and second-order models. Generation is more complicated – the algorithm I use for generation runs as follows:

1. Select random state:
    A

2. Select the next observation from the state transition matrix using state:
    $A \rightarrow B$

3. Append observation to previous state:
    A,B

4. Select next observation from state transition matrix using current state:
    $A, B \rightarrow A$

5. Append observation to previous state:
    A,B,A

6. Select next state from state transition matrix using current state:
    $A, B, A \rightarrow ?$

You will notice that we hit a snag at the end there – assuming we have only made first- and second-order matrices, there are no observations available following the state 'A,B,A'. At this point, I add the following:

1. If no observations are available, reduce the order of the previous state by one, from A,B,A to B,A

2. Select observations state from state transition matrix using B,A

3. Back to step 1 until you receive an observation

The reduction of order continues until observations are available. Earlier, I mentioned that sequence models can get stuck in a repetitive state loop, with only one possible next state. Consider the following matrix – for several states, there is only one next option, i.e. B:

	A	B
A,A	0	1
A,B	0.666	0.333
B,A	0	1
B,B	0	1

$$(20.7)$$

To prevent a sequence made from only Bs, I add an extra logical test to the algorithm: if there is *only one observation in the matrix* for this state, or there are no observations at all, reduce the order and sample again. This is the trick to avoiding repetitive loops. You only sample from rows in the matrix with at least two options.

I think that is enough abstract thinking about Markov models for now. I am keen to proceed with the implementation. If you want to read some analysis about sequence prediction performance and such for musical and other sequences using variable order Markov models, and you are ready for a technical and mathematical deep-dive, I refer you to [2].

20.7 Progress check

You should now have an insight into the research field of algorithmic composition and the idea of musical agents. You should also have an understanding of the construction of Markov models from simple letter sequences. You should be able to explain the idea of order in a model, states, observations and state transition matrices. You have also seen the variable order Markov model algorithm you will use in the following chapters to construct an improvising agent.

21

Programming with Markov models

In this chapter, I will show you how to use a variable-order Markov model library I have created. The library is written in C++. We will work on various command line experiments, feeding the model different sequences and investigating how it behaves.

21.1 Setting up the project

I am working on the assumption that you have followed the set-up instructions in the book's first part and have a working install of CMake and an IDE. If you still need to complete the set-up, you should return to that part of the book and ensure you have a working build environment on your machine. You should download the code pack that comes with the book. The project in the pack that you need to start from is described in the repository guide section 39.4.1. The project is quite simple – it has a main.cpp file, which just includes the MarkovModelCPP library header file and provides an empty main function. Open up the project in your IDE and attempt a test build to ensure everything is set up correctly.

21.2 Build a Markov model

Let's dive straight in and build some Markov models. Open the file main.cpp. You should see an empty main function in there. The following code sends a sequence of letters into a Markov model – add it to your main function and run it:

```
1  int main()
2  {
3      MarkovManager mm{};
4      mm.putEvent("A");
5      mm.putEvent("B");
6      mm.putEvent("A");
```

169

169

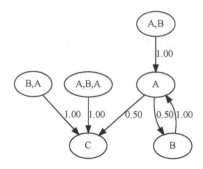

FIGURE 21.1
Example of the Markov model generated by some simple code.

```
7     mm.putEvent("C");
8     std::cout << mm.getModelAsString() << std::endl;
9 }
```

You should see some output like this:

```
1 1,A,:2,B,C,
2 1,B,:1,A,
3 2,A,B,:1,A,
4 2,B,A,:1,C,
5 3,A,B,A,:1,C,
```

The getModelAsString function generates a string representation of the current model. Considering what you know about Markov models, what do you think that output represents? Look carefully – the format of each line is:

number,state,:number,observations.

The first number signifies the order of the state. The state is a comma-separated list of the symbols representing the state. The second number is the number of observations seen to follow that state. The observations are a comma-separated list of observations. Try adding more events to the model and checking how the model string changes. Figure 21.1 illustrates the model as it stands after running that code. See if you can connect that diagram to your interpretation of the model's string representation.

You can pass any strings you like as the states for the model. Here is the signature for the putEvent function from MarkovManager.h:

```
1     void putEvent(state_single symbol);
```

'state_single' is the type for the argument. If you ctrl-click on that type in your IDE, you should see this from MarkovMode.h:

```
1 typedef std::string state_single;
```

This means that state_single is just a renamed std string type. So you can send in any strings you like for your symbols. How about something musical? We can add '#' to the letters to specify sharps and use '-' to represent a rest:

```
1 mm.putEvent("A");
2 mm.putEvent("B");
3 mm.putEvent("C#");
4 mm.putEvent("A");
5 mm.putEvent("B");
6 mm.putEvent("A");
7 mm.putEvent("C#");
8 mm.putEvent("-");
9 mm.putEvent("G#");
```

21.3   Generate from a Markov model

Now you have a model, you can use it to generate a new sequence. Here is a for loop that queries the model:

```
1 for (auto i=0;i<5;++i){
2     state_single next = mm.getEvent();
3     std::cout << "Next state " << next << std::endl;
4 }
```

I see output like this:

```
1 Next state B
2 Next state A
3 Next state C
4 Next state C
5 Next state C
```

You might see different outputs on different runs. This is a variable Markov model, which means it tries to find the highest-order output with an available observation following it. You can query the model for the order of the previous event using the getOrderOfLastEvent function:

```
1 for (auto i=0;i<5;++i){
2     state_single next = mm.getEvent();
3     int order = mm.getOrderOfLastEvent();
4     std::cout << "Next state " << next << " order " << order << std::
      endl;
5 }
```

I see output like this (which might be different each time):

```
1  Next state C order 0
2  Next state A order 0
3  Next state B order 1
4  Next state A order 2
5  Next state C order 3
```

What is going on here? Why does the order increase up to 3? The MarkovManager object maintains a memory of the state as the numbers are fed in. This allows it to build the state transition matrix. When it generates, it maintains a separate memory, which is the generation state, so the state it will use to query the state transition matrix. As the generation proceeds, the memory grows and can query with increasing order. You can try sending in the needChoices flag to getEvent. In the previous chapter, I explained how the Markov model can become repetitive or uninteresting if you allow it to sample from transition matrix rows with only one choice. If you set needChoices to false, it will be allowed to sample rows with only one observation. You will achieve higher order but less varied output.

21.4   Things to explore with Markov modelling

You can do many things with this command line Markov model code. For fun, try some experiments with words. You could read words from a text file, build a Markov model of the word sequence, and then generate new sequences in that style. You could experiment technically with the model and measure its ability to predict sequences. If you have trained on the words from a book, how often does it guess the next word correctly? What happens if you experiment with the maxOrder parameter for the MarkovManager constructor? This controls how many orders the model stores.

Another interesting thing to do is to create a chatbot that builds a model of the person it is talking to as the conversation proceeds. One of my first experiences with artificial intelligence was interacting with a similar chatbot on my Commodore Amiga computer in the early 1990s. A cover disk from an Amiga magazine contained several examples of chatbots, one of which worked like this. I don't know how sophisticated the underlying model was or what kind of model it used, but it certainly inspired me to think about artificial intelligence and fuelled some of my long journey to writing this book.

21.5   Progress check

You should have trained some Markov models using the MarkovModelCPP library. You have experimented with the needChoices parameter for the getEvent function on MarkovManager. You may have carried out some experiments to investigate the sequence modelling quality.

22

Starting the Improviser plugin

In this chapter, you will start building the Markov improviser plugin, which models incoming MIDI data and generates MIDI data back out from that model in a similar but 'interestingly different' style. The chapter starts with an overview of the plugin's main components, then explains how to prepare the JUCE project for building the plugin. Following that we dig into pitch modelling and find out how to access MIDI sent into the plugin, then to store it into a Markov model. You will see how you can extract data from the MIDI messages sent to the processBlock function, then convert them into a format that is suitable for the Markov model library we are using. At the end of the chapter you will have a basic monophonic improviser which can model then generate streams of notes.

22.1 Overview of the autonomous improviser plugin

The improviser plugin aims to model incoming MIDI data such that it can then generate MIDI data back out from that model. The MIDI data generated from the model should be patterned in a similar but 'interestingly different' style to the MIDI it has learned from. The improviser operates in real-time, updating its model all the time as MIDI data is received. This makes it exciting to interact with – it is constantly learning new tricks. Figure 22.1 illustrates the main components of the model. It models pitch, duration, onset times and velocity. These are the main elements that describe an incoming stream of MIDI notes: which notes? How long are the notes? When do the notes happen, and how loud are they? In the following chapters, you will see how to build models of these musical elements and combine them into a fully working improviser plugin.

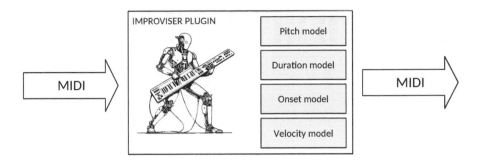

FIGURE 22.1
Overview of the autonomous improviser plugin. Yes, a keytar.

22.2   Preparing the JUCE project

Start by opening up the starter project for the Markov improviser plugin. It is the one described in section 39.4.2. I have configured it to pull in the Markov model library from a folder above using these lines in its CMakeLists.txt file:

```
1  add_library(markov-lib ../MarkovModelCPP/src/MarkovManager.cpp
2                         ../MarkovModelCPP/src/MarkovChain.cpp)
3  # ...
4  # AND then in the plugin section:
5  target_sources(midi-markov-v1
6  PRIVATE
7  src/PluginEditor.cpp
8  src/PluginProcessor.cpp
9  ../MarkovModelCPP/src/MarkovManager.cpp
10 ../MarkovModelCPP/src/MarkovChain.cpp
11 )
```

The project pulls the JUCE library from a folder two levels above, which will work if you download the whole code pack. This line in CMakeLists.txt specifies the location for JUCE:

```
1  add_subdirectory(../../JUCE ./JUCE)
```

The first reference to JUCE in that line should point to the location of the full JUCE distribution on your system. The CMakeLists.txt file also specifies a command line test project wherein you can quickly try things out with the Markov model library. These lines set it up:

```
1  add_executable(markov-expts src/MarkovExperiments.cpp)
```

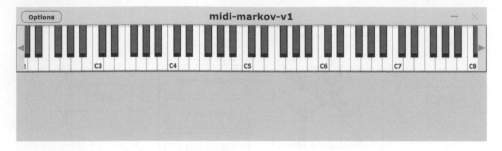

FIGURE 22.2
The user interface for the basic JUCE MIDI processing plugin.

```
2  target_link_libraries(markov-expts  markov-lib)
```

Compiling the whole plugin project every time to try something out with the underlying Markov library slows my development cycle. The markov-expts target builds very quickly as it is separated from the build for the Markov model library and the JUCE code, so you can use it to test and prototype things that do not require the complete JUCE program.

Going back to the JUCE target in this project, here are some more things that are useful to know:

1. The project is a JUCE plugin project, but you can also run it in Standalone mode

2. The plugin target has 'IS_MIDI_EFFECT', 'NEEDS_MIDI_INPUT' and 'NEEDS_MIDI_OUTPUT' set to TRUE as it will be a MIDI effect-type plugin

3. You should change the settings for COMPANY_NAME, PLUGIN_MANUFACTURER_CODE and PRODUCT_NAME as you see fit

Once you have the starter project downloaded, unpacked and configured, generate the IDE project in your usual way, open it up in your usual IDE and build it.

22.3   Modelling pitch

We are going to start by modelling pitch sequences. The starter project already has a basic JUCE plugin architecture in place, including a handy on-screen piano keyboard. Figure 22.2 illustrates the user interface for the starter project. The on-screen keyboard can generate MIDI messages, which you can use to create data for the Markov models.

Take a look at the user interface code file PluginEditor.cpp. You will see two functions there called handleNoteOn and handleNoteOff. In terms of C++ techniques, the MidiMarkovEditor class extends the juce-MidiKeyboardState-Listener class. That class is abstract – it has no implementation for those two functions. Therefore, any class that extends it must implement those two functions. Those functions will be called by the piano keyboard widget when the user clicks and releases the notes. So, if you want to print out the notes as they come in, you can edit the handleNoteOn function as follows:

```
juce::MidiMessage msg1 =
    juce::MidiMessage::noteOn(midiChannel,
                             midiNoteNumber,
                             velocity);
// add this line
DBG(msg1.getNoteNumber());
```

To see debug output on Windows/ Visual Studio, you must enable the Immediate mode panel as described in the set-up section 3.5.1. Build and run the plugin in Standalone mode, then click on the keyboard. You should see output something like this:

```
77
78
79
80
81
82
```

22.3.1 Storing the MIDI messages for later processing

You may have spotted that the handleNoteOn function calls audioProcessor.addMidi, sending it the MIDI message. This completes the flow for these MIDI messages from the on-screen keyboard via the plugin's UI over to the plugin's audio processor. But what should the audio processor do with the MIDI message? Ultimately, it will be adding them to the Markov model, but we need to consider how to do that.

First, the user can click the keyboard anytime, triggering that addMidi function. I prefer to choose when I want to handle that MIDI, so I will store it for later processing. As a general principle – do not let the user mess with your audio processor – only do it on your terms when you want to! You can store MIDI messages using a JUCE MidiBuffer object. Add this to the private section in PluginProcessor.h:

```
juce::MidiBuffer midiToProcess;
```

Then, in PluginProcessor.cpp, edit the addMidi function so it adds the messages to the buffer 'midiToProcess':

```
1  midiToProcess.addEvent(msg, sampleOffset);
```

The sample offset is zero for now – that is passed in from the user interface when it calls addMidi. So now the audio processor is stashing the MIDI messages away for later processing.

22.4   Building a Markov model from the MIDI notes

Let's see if you can deal with those stored MIDI messages by building a Markov model of the note sequence. To make a Markov model, you will use the Markov-Manager class. The best place to locate the MarkovManager object in the application is in the audio processor class defined in PluginProcessor.h. This is for a few reasons:

1. Most importantly, only the audio processor has access to sample-accurate timing data, allowing it to precisely represent time in the Markov model

2. The audio processor object persists for the lifetime of the plugin, whereas the user interface can be killed and re-created

3. Whilst the on-screen MIDI keyboard talks to the plugin user interface, the audio processor receives the most important MIDI in its process-Block function, the MIDI from the plugin host

Add the MarkovManager object to the private section of the audio processor class in PluginProcessor.h, along with the include:

```
1  // at the top - assumes you have MM two folders up
2  #include "../../MarkovModelCPP/src/MarkovManager.h"
3  // ...
4  // in the private fields of the MidiMarkovProcessor class:
5  MarkovManager pitchModel;
```

In PluginProcessor.cpp, initialise the pitchModel variable in the initialiser list on the constructor:

```
1  ...
2  // just above the body of the constructor
3  , pitchModel{}
4  {
5      // the body of the constructor here
6  }
```

We will now add any notes we have received to the pitchModel object using its putEvent function. In the MidiMarkovProcessor::processBlock function, add the following code:

```
1  if (midiToProcess.getNumEvents() > 0){
2      midiMessages.addEvents(midiToProcess,
3                              midiToProcess.getFirstEventTime(),
4                              midiToProcess.getLastEventTime()+1, 0);
5      midiToProcess.clear();
6  }
```

This code checks if the MIDI buffer midiToProcess has any events (which it will if the user clicks on the on-screen MIDI keyboard). If it does, it grabs the events using a time window spanning the earliest to the latest event. Then it calls addEvents on midiMessages, which is the buffer of MIDI messages sent to processBlock – remember it has this function prototype:

```
1  void processBlock (juce::AudioBuffer<float>& audioBuffer,
2                      juce::MidiBuffer& midiMessages) override;
```

The second argument will contain any MIDI messages sent from the plugin host (if you run the plugin in a DAW or similar). So, we are adding all the messages collected from the on-screen MIDI keyboard to the messages received from the host. All the MIDI we need to deal with is now in one place.

After all that, we remove the messages stored from the on-screen keyboard in midiToProcess by calling clear. We are now ready to push the notes to the Markov model. In processBlock, iterate over the midiMessages and add all note-on messages to the Markov model:

```
1  for (const auto metadata : midiMessages){
2      auto message = metadata.getMessage();
3      if (message.isNoteOn()){
4          pitchModel.putEvent(std::to_string(message.getNoteNumber()));
5          DBG("Markov model: " << pitchModel.getModelAsString());
6      }
7  }
```

This code pulls any MIDI note on messages out of the buffer and adds them to the Markov model. Then, it prints out the full state of the model. Try running this and pressing the keys on the on-screen piano keyboard. You should see output like this, showing that your model is rapidly growing:

```
1  9,96,96,95,77,79,81,89,91,93,:1,95,
2  9,96,96,95,93,91,89,86,84,82,:1,80,
```

If you have a MIDI controller keyboard handy, you can run the application in Standalone mode and configure the keyboard as an input device. You can then put notes into your model by playing on the keys.

22.5   Generating MIDI notes from the model

Now you have a model, you can use it to generate notes and send them to the output. You can start by calling getEvent on the model and printing the note. Here is some code to do that:

```
for (const auto metadata : midiMessages){
    auto message = metadata.getMessage();
    if (message.isNoteOn()){
        pitchModel.putEvent(std::to_string(message.getNoteNumber()));
        int note = std::stoi(pitchModel.getEvent(true));
        DBG( "incoming note:"
                << message.getNoteNumber()
                << " model note:"
                << note
                << " order "
                << pitchModel.getOrderOfLastEvent()
                << std::endl);
    }
}
```

When I run the plugin in Standalone mode and press a few buttons on the piano keyboard, I see an output like this:

```
incoming note:89 model note:82 order 0
incoming note:86 model note:84 order 0
incoming note:71 model note:89 order 1
incoming note:72 model note:82 order 0
incoming note:81 model note:89 order 0
incoming note:74 model note:81 order 0
incoming note:88 model note:76 order 1
incoming note:79 model note:91 order 1
```

Here, it is crucial to understand the design of the variable Markov model provided in my implementation. If you pass the 'true' parameter to the getEvent function, it will choose an order for the model with at least two choices for the next state. The model remembers the order of the last state it used for look-up, and you can query this with the getOrderOfLastEvent function. If you have only passed a few MIDI notes (events) to the model, it will generally generate at a low order because it has not seen higher orders and certainly not with different outcomes. In my example above, the Markov model mainly operates in zero order (sampling from all observations, ignoring state) or first order. As you put more events in, it will be able to sample at a higher order – the model is getting richer. Experiment with sending note sequences into the model with the on-screen keyboard or with your MIDI keyboard. If you play in repeated patterns, does it increase its order? Try some variations. Training a Markov model during a live performance is like playing an instrument – you must learn how to get what you want out of it through practice.

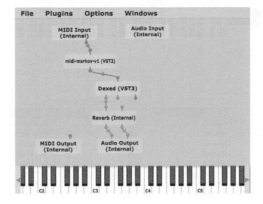

FIGURE 22.3
The MIDI Markov plugin running in AudioPluginHost. Note how it receives MIDI
and then passes it on to the Dexed synthesiser.

22.6 Sending MIDI notes from the Markov plugin to another plugin

The improviser plugin is intended to be a MIDI effect, so a typical set-up is that
it sits in between a source of MIDI data and a synthesiser that can play that data.
Figure 22.3 illustrates this set-up, with the MIDI Markov plugin processing MIDI
before it goes to the Dexed synthesiser. To make your plugin send a note from
its output; you just need to add that note to the midiMessages buffer sent to the
plugin's processBlock function. Think of it like a train arriving at a station. The
train arrives in your plugin; you process any messages that come on the train;
you then add any messages you want to send to the following plugin. The train
leaves your plugin and goes to the next one in the chain. The following code shows
you how to read a note from the pitchModel and then add it to the midiMessages
buffer:

```
// get a note from the mode, converting to an int
int note = std::stoi(pitchModel.getEvent(true));
// convert the int into a note on message
juce::MidiMessage nOn = juce::MidiMessage::noteOn(1,
                                                  note,
                                                  0.5f);
// add it to the outgoing buffer
midiMessages.addEvent(nOn, 0);
```

If you call that directly in processBlock, you will send a message every time
processBlock is called. That is probably a bad idea as processBlock gets called

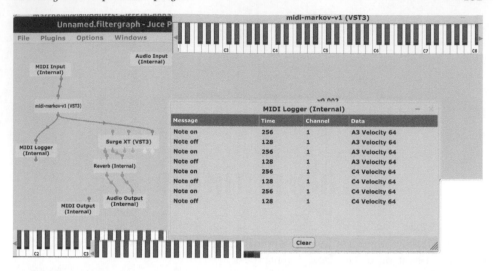

FIGURE 22.4

Using AudioPluginHost's MIDI Logger plugin to observe the MIDI coming out of the Markov plugin.

sampleRate / blocksize times per second, for example 43 times a second for a typical 44.1k/1024 set-up. So it is time to make a design decision – when should the music improviser play a note? Come up with your own idea and implement it. The most straightforward idea is to play a note when it receives a note. To do that, you could add this code to the end of processBlock after you have added any incoming notes to the Markov model:

```
if (midiMessages.getNumEvents() > 0){
    int note = std::stoi(pitchModel.getEvent(true));
    juce::MidiMessage nOn = juce::MidiMessage::noteOn(1,
                                                       note,
                                                       0.5f);
    midiMessages.addEvent(nOn, 0);
}
```

Try it out – you should see notes coming out of the plugin when it receives notes. You will find that the plugin generates notes when it receives note-on and note-off messages. The on-screen keyboard generates a note-off message when you release the mouse after pressing one of the piano keys. Can you think of a way to block the double note trigger?

You can test the plugin with the AudioPluginHost, as shown in figure 22.4. This figure also shows how you can wire up the MIDI Logger plugin to print out any MIDI messages you generate from the Markov plugin. The MIDI Logger plugin comes as a built-in plugin with AudioPluginHost.

22.6.1 Separate MIDI buffer for generated notes

We are going to make that processBlock code a little more complex now. The problem with storing the generated MIDI notes from the model straight onto the outgoing MIDI buffer is that we cannot choose to wipe the buffer before sending our notes. If you do not want the improviser to parrot the incoming notes back out of its output, you need to wipe those notes and then replace them with your generated notes. To do that, you can create a temporary MIDI buffer in processBlock and write the generated notes there:

```
1  // add this:
2  juce::MidiBuffer generatedMessages;
3  if (midiMessages.getNumEvents() > 0){
4      int note = std::stoi(pitchModel.getEvent(true));
5      juce::MidiMessage nOn = juce::MidiMessage::noteOn(1,
6                                                        note,
7                                                        0.5f);
8      // add the messages to the temp buffer
9      generatedMessages.addEvent(nOn, 0);
10 }
11 // now you can clear the outgoing buffer if you want
12 midiMessages.clear();
13 // then add your generated messages
14 midiMessages.addEvents(generatedMessages, generatedMessages.
       getFirstEventTime(), -1, 0);
```

22.7 Note-off messages

You will notice in your testing that the notes never end. This is because you must send MIDI note-off messages to tell the synthesiser to stop playing the notes. This is complex because you want to send the note-off messages at a different time than you send the note-on because sending note-on and note-off at the same time will result in very short notes, or the synthesiser might even ignore the note-off messages. You need to send the note-off messages a little later, say, one second later. Since processBlock only gives you a small window of time to send your messages, you will need to wait for a later call to processBlock to generate the note-off message. So you need some way to remember between calls to processBlock that you have some note-off messages to send.

A simple way to achieve this, and this will help you later when you work on modelling time, is to have an array in which you remember when you need to send note-offs. The array will have 127 elements, one for each possible note. Another variable will allow you to keep track of the number of elapsedSamples.

```
1  ... put this after your note processing code
2  for (auto i=0;i<127;++i){
3      if (noteOffTimes[i] > 0 &&
4          noteOffTimes[i] < elapsedSamples){
5          juce::MidiMessage nOff = juce::MidiMessage::noteOff(1, i, 0.0
   f);
6          generatedMessages.addEvent(nOff, 0);
7          noteOffTimes[i] = 0;
8      }
9  }
```

The unsigned long type is appropriate as the number of elapsed samples might become quite large. Add these to the private section in processBlock.h:

```
1  unsigned long noteOffTimes[127];
2  unsigned long elapsedSamples;
```

Then, initialise the variables in the constructor:

```
1  elapsedSamples = 0;// or if you are a real engineer,
2                     // do this C++11-style in the initialiser list
3  for (auto i=0;i<127;++i){
4    noteOffTimes[i] = 0;
5  }
```

First of all, let's deal with managing elapsedSamples. At the end of the processBlock function, just add the length of the block to the elapsedSamples:

```
1  elapsedSamples += buffer.getNumSamples();
```

That means every time processBlock is called, elapsedSamples increases by the number of samples in the block, e.g. 1024. When you want to remember to play a note off for later, store the time at which the note off should be triggered in the noteOffTimes. In the processBlock's note-generating code:

```
1  ...
2  int note = std::stoi(pitchModel.getEvent(true));
3  juce::MidiMessage nOn = juce::MidiMessage::noteOn(1, note, 0.5f);
4  generatedMessages.addEvent(nOn, 0);
5  // now the note off
6  noteOffTimes[note] = elapsedSamples + getSampleRate();
```

I set the note-off time to elapsedSamples + getSampleRate(), which equates to 'one second in the future'. Now you need to check the noteOffTimes array every time processBlock is called and add note-off messages if necessary:

In that code, I check if elapsedSamples has progressed beyond the time at which the note-off should occur. If so, it is time to send the note off. Send it and reset the relevant value in noteOffTimes. Test this out in the AudioPluginHost with the MIDI logger. You should see MIDI note-on and MIDI note-off messages coming out of your plugin. You need to quit and restart the AudioPluginHost if you want it to load the latest version of your plugin. If you save the layout, it should reload automatically next time, making things easier.

This is not sample-accurate timing as you do not check the timings for every sample in the buffer, just once per buffer. Feel free to work up a more accurate timing system, but this will work for now, and the worst timing will be a buffer length out.

22.8 Add a reset button

Now is a good time to add a reset button to the Markov plugin user interface, which resets the Markov model. Add a user interface TextButton object to the PluginEditor.h:

```
juce::TextButton resetButton;
```

Call addAndMakeVisible in the constructor in PluginEditor.cpp:

```
addAndMakeVisible(resetButton);
```

Set the bounds in resize:

```
...
miniPianoKbd.setBounds(0, rowHeight*row, getWidth(), rowHeight);
row ++ ;
resetButton.setBounds(0, rowHeight*row, getWidth(), rowHeight);
```

Compile and verify you can see your button. Then verify that you have implemented the button listener interface in PluginEditor.h, which the template project already does. Then add a function prototype to the class in PluginProcessor.h to allow the UI to call resetModel:

```
void resetMarkovModel();
```

Then, an implementation in PluginProcessor.cpp:

```
void MidiMarkovProcessor::resetMarkovModel()
{
  pitchModel.reset();
}
```

Finally, put some code in PluginEditor.cpp to deal with the reset:

```
void MidiMarkovEditor::buttonClicked(juce::Button* btn)
{
    if (btn == &resetButton){
        audioProcessor.resetMarkovModel();
    }
}
```

22.9 Progress check

At this point, you should have a fully functional, monophonic Markov model plugin. You should be able to load up the plugin in your DAW or other plugin host and see it receive, model and generate MIDI. You can see an example of the results of this chapter in code example 39.4.3.

23

Modelling note onset times

In this chapter, you will add time modelling to the improviser. You will do this using two separate Markov chains: one for note duration and one for 'inter-onset-interval', which is the time that elapses between the start of notes. You learn how the processBlock function provides you with MIDI messages with sample-accurate timing data. You will use this timing data to measure note durations and inter-onset intervals. You can then create models of the two aspects of time, which you can use to control the behaviour of the note-generating model.

23.1   Note duration and inter-onset-interval

We are now going to teach the musical agent about timing. Timing is very important in music; I have been fortunate to work with musicians with an excellent time sense. Knowing when to stop playing is another problem entirely. We will consider two facets of timing: how long to play the notes for and how long to wait between notes. I will refer to these two as note duration and inter-onset interval (IOI). Figure 23.1 illustrates these properties of time on a typical piano-roll view. You

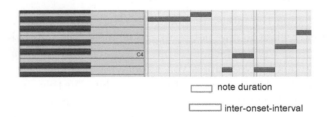

FIGURE 23.1
Note duration is the length the note plays for. Inter-onset interval is the time that elapses between the start of consecutive notes.

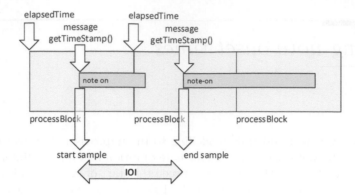

FIGURE 23.2
Measuring inter-onset-intervals. The IOI is the number of samples between the start and end sample. elapsedSamples is the absolute number of elapsed samples since the program started and is updated every time processBlock is called; message.getTimestamp() is the offset of the message in samples within the current block.

can continue working on the plugin code you had at the end of the last chapter, or if you want a clean start, you can start with the project from section 39.4.3 in the repo guide.

23.2 Model the inter-onset-interval

We will start by modelling the inter-onset intervals (IOI) because this will allow the improviser to decide when to play notes. At the moment, the improviser plays a note whenever you do. It will be much more interesting if it plays notes according to a model of the intervals between notes you are playing. For now, we can stick with fixed duration notes; we will model note duration later in this chapter. The first step in measuring IOIs is capturing the absolute time a note begins, i.e. the start sample value shown in figure 23.2. We do not need to worry about which notes are being played – the IOI measures the interval between any notes played. Let's add variables to store the IOI Markov model and the start sample value to the PluginProcessor.h private section:

```
1  MarkovManager iOIModel;
```

```
2  unsigned long lastNoteOnTime;
```

Make sure you call reset on that model in your resetMarkovModel function in PluginProcessor.cpp. Then initialise lastNoteOnTime to zero in the initialiser list and setup the iOIModel:

```
1      ,pitchModel{}, iOIModel{}, lastNoteOnTime{0}, elapsedSamples{0}
2      //constructor body here
3      {
```

Now, when we receive a MIDI note-on message in processBlock, store the absolute time in samples at which it occurred (elapsedTime + message.getTimestamp()) to lastNoteOnTime:

```
1  for (const auto metadata : midiMessages)
2  {
3      auto message = metadata.getMessage();
4      if (message.isNoteOn())
5      {
6          unsigned long exactNoteOnTime = elapsedSamples + message.
   getTimeStamp();
7          lastNoteOnTime = exactNoteOnTime;
8          DBG("Note on at: " << exactNoteOnTime);
9      }
10 }
```

I see output like this from that code – make sure you see something similar when you run the app in standalone mode and click on the on-screen keyboard:

```
1  Note on at: 55296
2  Note on at: 82944
3  Note on at: 110592
4  Note on at: 152064
5  Note on at: 177664
```

Before things get any more complex, you should modularise the code. You can add a function called analyseIoI to PluginProcessor.cpp/h

```
1  // in the header, private section:
2  void analyseIoI(const juce::MidiBuffer& midiMessages);
3  // in the cpp, add the code from above:
4  void MidiMarkovProcessor::analyseIoI(const juce::MidiBuffer&
      midiMessages)
5  {
6    for (const auto metadata : midiMessages)
7    {
8      ...
9    }
10 }
11 // then, in processBlock, call the function:
12 updateNoteOnTime(midiMessages);
```

Note how I used const and &. const means we will not modify the sent data (the

MidiBuffer) – we are just reading it. The ampersand means the data will be sent by reference instead of sending in a copy. This makes the code more efficient as it does not need to copy the MidiBuffer every time that function gets called. Whilst we are being good engineers, you can also modularise the pitch modelling code from processBlock into its own function if you like. I call my pitch management function analysePitches:

```
1 // in the header, private section:
2 void analysePitches(const juce::MidiBuffer& midiMessages);
3 // in the cpp, add the pitch modelling code:
4 void MidiMarkovProcessor::analysePitches(const juce::MidiBuffer&
      midiMessages)
5 {
6   for (const auto metadata : midiMessages){
7     auto message = metadata.getMessage();
8     if (message.isNoteOn()){
9       pitchModel.putEvent(std::to_string(message.getNoteNumber()));
10    }
11  }
12 }
```

Now you need to update the analyseIoI function, so it computes the IOI and adds it to the IOI Markov model:

```
1 unsigned long exactNoteOnTime = elapsedSamples + message.getTimeStamp
      ();
2 // compute the IOI
3 unsigned long iOI = exactNoteOnTime \textendash{} lastNoteOnTime;
4 // add it to the model
5 iOIModel.putEvent(std::to_string(iOI));
6 lastNoteOnTime = exactNoteOnTime;
7 DBG("Note on at: " << exactNoteOnTime << " IOI " << iOI);
```

Now build and run in Standalone mode. Tap keys on the on-screen keyboard and monitor the IOIs detected. Tapping at approximately one-second intervals, I see the following output (don't tell anyone that my timing is this bad – I sometimes claim to be a drummer!):

```
1 Note on at: 778752 IOI 45568
2 Note on at: 820224 IOI 41472
3 Note on at: 860672 IOI 40448
```

My experience with improvisers is that you do not want them to model excessively long pauses. This can lead to long pauses in the output, and you will be wondering if things are working correctly if that happens during your performance. With some checking code, you can only add IOIs below a certain length to the model. Something like this will do the trick – only add IOIs less than 2 seconds and more than some short time to avoid re-triggers:

```
1 if (iOI < getSampleRate() * 2 &&
2     iOI > getSampleRate() * 0.05){
```

```
3    iOIModel.putEvent(std::to_string(iOI));
4  }
```

23.3 Generate with the inter-onset-intervals

You are making significant progress – you have a model of the inter-onset intervals, but how can you use it to specify the wait between notes? At the moment, the model simply plays a note whenever it receives a note. This could be more musically interesting. Instead, we want the model to pull an IOI value from the IOI model and to wait that long before it plays the following note.

23.3.1 Refactor the note generator

Before we start editing the note-generating logic in processBlock, let's modularise that code into its own function. Take the code you have in processBlock that sends messages out from the pitch model and re-organise it like this in PluginProcessor.h and cpp:

```
1  // in the header, private section:
2  juce::MidiBuffer generateNotesFromModel(const juce::MidiBuffer&
       incomingMessages);
3
4  // in the cpp:
5  juce::MidiBuffer MidiMarkovProcessor::generateNotesFromModel(
6      const juce::MidiBuffer& incomingNotes)
7  {
8    juce::MidiBuffer generatedMessages{};
9    // add notes from the model to the temp buffer
10   if (incomingNotes.getNumEvents() > 0){
11     int note = std::stoi(pitchModel.getEvent(true));
12     juce::MidiMessage nOn = juce::MidiMessage::noteOn(1,
13                                                       note,
14                                                       0.5f);
15     // add the messages to the temp buffer
16     generatedMessages.addEvent(nOn, 0);
17     noteOffTimes[note] = elapsedSamples + getSampleRate();
18   }
19   return generatedMessages;
20 }
```

Then you can just call it like this in processBlock:

```
1  juce::MidiBuffer generatedMessages = generateNotesFromModel(
       midiMessages);
```

Build and run to check things are working as you expect. If you run the plugin inside AudioPluginHost, you should see MIDI coming from the plugin's output when you send notes into it, as before.

23.3.2   Decide when to play notes

Now that things look tidier let's deal with decision-making using the IOI model. You will need two functions and a variable to make this work. Add these to your PluginProcessor.h:

```
1  // the time when the next note should be played
2  unsigned long modelPlayNoteTime;
3  // return true if it is time to play a note
4  bool isTimeToPlayNote(unsigned long currentTime);
5  // call after playing a note
6  void updateTimeForNextPlay();
```

Set modelPlayNoteTime to zero in the constructor or initialiser list. The isTimeToPlayNote function works like this – it checks if the current time has passed the time to play the next note and returns true if it has:

```
1  bool MidiMarkovProcessor::isTimeToPlayNote(unsigned long currentTime)
2  {
3    if (currentTime >= modelPlayNoteTime){
4      return true;// time to play a note
5    }
6    else {
7      return false;
8    }
9  }
```

Now you can update your generateNotesFromModel function so it uses that function to decide if it should play a note – in PluginProcessor.cpp/ generateNotes-FromModel:

```
1  if (isTimeToPlayNote(elapsedSamples)){
2      // add a note-on to generatedMessages
3      ...
4      // then set the note-off time for that note in noteOffTimes
5      ...
6      // finally, query the IOI model for the next note on time
7      unsigned long nextIoI = std::stoi(iOIModel.getEvent());
8      if (nextIoI > 0){// sanity check the model output
9          modelPlayNoteTime = elapsedSamples + nextIoI;
10     }
11 }
```

I found that this code endlessly sends notes until the plugin receives some MIDI notes. I will let you figure out how to solve that bootstrapping problem. A clue is that you can have a boolean that tells you if you are in the startup state or have received MIDI. It starts as true, then once you've received some MIDI, set it to false. If it is true, do not send out any MIDI notes. There was a lot of logic and refactoring in this section. Another aspect you can work on for yourself is perfecting the timing of sending the note-on and note-off messages. At the moment, the messages are always triggered at the start of the first audio block occurring after their trigger time. This means a note could be 'block-length' samples late. The solution is to find all notes that should be triggered between elapsedSamples and elapsedSamples + block size. Then, set timestamp offsets on those messages according to where they should fall in the block. I leave it to you to figure that out.

23.4 Progress check

You should have a plugin that can receive MIDI notes and analyse monophonic pitch sequences and inter-onset intervals. If you want to see a working implementation of the improviser up to this point, I refer you to example 39.4.4 in the repo guide.

24

Modelling note duration

In this chapter, you will complete the modelling of time and rhythm in the improviser by implementing a note duration model. The note duration model remembers when different notes start and when they end. It then computes the duration of the notes and models this using a separate Markov model. At the end of the chapter, you should have a reasonable plugin working that can improviser with monophonic MIDI sequences.

24.1   Calculating note durations

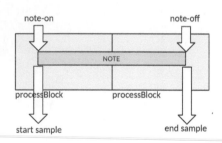

FIGURE 24.1
Measuring note duration has to cope with notes that fall across multiple calls to processBlock.

At the moment, the improviser always plays one-second-long notes. Ideally, we would like to set the note durations using a model of the incoming note durations, as we do for pitch and IOI. To measure note duration, you can use a similar technique to the one used to prepare note-off messages in the monophonic pitch model. We managed note-off times by working out the absolute time in elapsed samples at which the note-off should be sent and storing it into an array with a slot for each of the 127 possible notes. Each time processBlock ran, we checked in the array to see if the current time had passed the time at which any of the note-offs should be sent. If a note-off needed to be sent, we added it to the outgoing MIDI message buffer.

Returning to note duration, we will measure note duration by storing the starting sample for incoming note-ons in an array with a slot for each note. When the note-off for that note arrives, we can compute the number of samples that have elapsed since the stored note-on time. That will be the duration of the note.

Figure 24.1 illustrates the concept of measuring the number of samples between note-on and note-off. It does not show the array, but there would be a separate 'start sample' for each possible note. It does show how notes can fall across calls to processBlock.

Right, it's time to edit some code. Add a new private data member in Plugin-Processor.h:

```
unsigned long noteOnTimes[127];
```

Initialise all values to zero in the constructor, as you already did for the note-OffTimes array; in fact, just use the same loop:

```
for (auto i=0;i<127;++i){
    noteOffTimes[i] = 0;
    noteOnTimes[i] = 0;
}
```

Then, when you detect a note in your analysePitches function, log elapsed time into the array for that note:

```
...
for (const auto metadata : midiMessages){
    auto message = metadata.getMessage();
    if (message.isNoteOn()){
        DBG("Msg timestamp " << message.getTimeStamp());
        pitchModel.putEvent(std::to_string(message.getNoteNumber()));
        noteOnTimes[message.getNoteNumber()]
                = elapsedSamples + message.getTimeStamp();
...
```

Notice that I did not just use elapsedSamples – elapsedSamples only updates once per call to processBlock, but the note might have occurred within that block. On my system running with a block size of 2048, I see output like this from that code when I run the plugin in the AudioPluginHost as shown in 24.2.

```
Msg timestamp 1107
Msg timestamp 593
Msg timestamp 1133
Msg timestamp 399
Msg timestamp 1775
Msg timestamp 102
Msg timestamp 623
```

So now you know when any incoming notes started. The next step is to detect when the notes end and then compute their length. Going back to the note iterator in analysePitches, add a new block to deal with note-offs:

```
if (message.isNoteOff()){
```

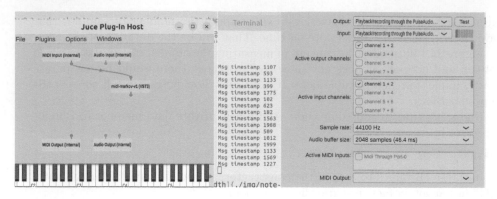

FIGURE 24.2

Testing the getTimestamp function on note–on messages – the timestamp is always between zero and the block size of 2048.

```
2    unsigned long noteOffTime = elapsedSamples + message.getTimeStamp
     ();
3    unsigned long noteLength = noteOffTime -
4                          noteOnTimes[message.getNoteNumber()];
5    DBG( "n: " << message.getNoteNumber()
6              << " lasted "
7              << noteLength);
8  }
```

Try it! I see output like this when I play notes on the plugin's built-in piano keyboard:

```
1  n: 65 lasted 36555
2  n: 72 lasted 37940
3  n: 72 lasted 4880
4  n: 72 lasted 1952
5  n: 72 lasted 7587
```

Try holding down the notes for longer and verifying that the lengths of the notes look correct in the printout. Now, you need a Markov model to model these note durations. You can simply add another MarkovManager object to your PluginProcessor.h private fields:

```
1  MarkovManager noteDurationModel;
```

Now call putEvent on that model in your note iterator code in analysePitches:

```
1  if (message.isNoteOff()){
2    unsigned long noteOffTime = elapsedSamples + message.getTimeStamp
     ();
3    unsigned long noteLength = noteOffTime -
4                          noteOnTimes[message.getNoteNumber()];
5    noteDurationModel.putEvent(std::to_string(noteLength));
```

```
6
7 }
```

Engineers like modular code; analysePitches is supposed to deal with note pitches, not durations. Let's create a separate function called analyseDurations that contains the note duration code, leaving the pitch code in analysePitches. Something like this:

```
1 void MidiMarkovProcessor::analysePitches(
2     const juce::MidiBuffer& midiMessages)
3 {
4   for (const auto metadata : midiMessages)
5   {
6     auto message = metadata.getMessage();
7     if (message.isNoteOn())
8     {
9       pitchModel.putEvent(std::to_string(message.getNoteNumber()));
10     }
11   }
12 }
```

And this:

```
1 void MidiMarkovProcessor::analyseDuration(
2     const juce::MidiBuffer& midiMessages)
3 {
4     for (const auto metadata : midiMessages)
5     {
6         auto message = metadata.getMessage();
7         if (message.isNoteOn())
8         {
9             noteOnTimes[message.getNoteNumber()] = elapsedSamples +
    message.getTimeStamp();
10         }
11         if (message.isNoteOff()){
12             unsigned long noteOffTime =
13                     elapsedSamples + message.getTimeStamp();
14             unsigned long noteLength = noteOffTime -
15                                     noteOnTimes[message.getNoteNumber
    ()];
16             noteDurationModel.putEvent(std::to_string(noteLength));
17         }
18     }
19 }
```

Then your processBlock can contain three neat function calls like this:

```
1 analysePitches(midiMessages);
2 analyseDuration(midiMessages);
3 analyseIoI(midiMessages);
```

That feels better. Nothing worse than poor modularisation. Remember to reset the new model in the resetModel function on your PluginProcessor.

24.1.1 Design decisions for the improviser

I would like to pause to note that I have made a design decision for my improviser. I have decided to model note durations independently from note pitches. They are in two separate models. I have chosen to set things up this way because it will allow the model to associate duration patterns with note patterns that were not necessarily associated with the actual data. For me, this idea has creative merit. But it could be more technically perfect sequence modelling; better sequence modelling would create a compound state combining note duration and note pitch (and anything else we model later). What do you think? Once the improviser is more complete, you can experiment with these ideas.

24.2 Generating note durations from the model

Now you have a model of the note durations, how can you use it to dictate the duration of the notes played by the improviser? It is easier than working with IOI data! The answer is to use the model to decide how far the note-offs will occur in the future instead of hard coding the value. This is how we have been setting note-off times so far (this code should be in generateNotesFromModel):

```
1  // get a pitch from the pitch model
2  int note = std::stoi(pitchModel.getEvent(true));
3  juce::MidiMessage nOn = juce::MidiMessage::noteOn(1,
4                                                     note,
5                                                     0.5f);
6  generatedMessages.addEvent(nOn, 0);
7  // set the note-off time to one second in the future
8  noteOffTimes[note] = elapsedSamples + getSampleRate();
```

The last line dictates the length of the note, which is hardcoded to one second. Instead, try this line, which queries the note duration model for a length:

```
1  unsigned int duration = std::stoi(noteDurationModel.getEvent(true));
2  noteOffTimes[note] = elapsedSamples + duration;
```

Experiment with the plugin – send sequences of long and short notes to see if it is correctly modelling note duration. Use the AudioPluginHost MIDI Logger to check the raw MIDI output. You can see a fully working version of the plugin with monophonic pitch, IOI and duration modelling in the repo guide section 39.4.5.

24.3   Quantisation

If you send the improviser MIDI data from your DAW, it will probably already be quantised, meaning that the timing for the notes snaps to a timing grid. The improviser will also use this quantised timing for its models. But the timing will not be quantised if you are playing freely into the plugin. You need to quantise the timing to avoid ending up with a primarily low-order model. The reason is that each IOI and note duration will be different because it can fall at any point, sample-wise.

Can you work out a way to quantise the note durations and IOIs? If you quantise them, you will get a richer (higher-order) model as there will be more instances of each duration. Of course, increasing quantisation also makes the timing more 'robotic', and you might not want robotic timing. I leave it to you to experiment with timing.

24.4   Progress check

At this point, you should have a fully working Markov model-based musical agent plugin. It can learn in real-time from a live MIDI feed and generate its output at the same time as learning. You should also be able to think more about design decisions in the improviser. Are separate models for pitch, IOI and duration a good idea? Or will a compound model produce better results for you. Also, you should have thought about quantisation. Plenty for you to explore there.

25

Polyphonic Markov model

In this chapter, you will develop the Markov modelling improviser plugin further so that it can model polyphonic MIDI sequences. The basic principle here is that polyphonic states are just another type of state. Our first job will be to figure out how to pass polyphonic states to the Markov model. Then, we must work out how to process those states when generating from the model. You will also deal with some practicalities relating to humans playing chords on keyboards, where the note-ons do not happen at precisely the same time. The solution is a ChordDetector class, which provides you with chords according to some constraints about note-on timings.

25.1   Preparing polyphonic states

You can continue with the code you have from the previous chapter, or if you want a clean start, you can start from example 39.4.5, which contains most of the functionality to this point. We will get straight into the code now.

So far, you have been passing single note numbers to the Markov model like this:

```
markovModel.putEvent(
        std::to_string(message.getNoteNumber())
);
```

The MarkovManager putEvent function expects to receive a string, and the code above simply converts a MIDI note number into a string. So, if you want to send it a chord, you need to figure out how to encode several numbers into a string. Then, when the encoded numbers return from 'getEvent', you need code to convert back to several note numbers from your encoded string. Add the following functions to the private section of the PluginProcessor.h file:

```
std::string notesToMarkovState (const std::vector<int>& notesVec);
std::vector<int> markovStateToNotes (const std::string& notesStr);
```

These helper functions will allow you to convert between vectors of notes and strings. Here are the implementations:

200

```
1  std::string MidiMarkovProcessor::notesToMarkovState(
2                  const std::vector<int>& notesVec)
3  {
4  std::string state{""};
5  for (const int& note : notesVec){
6    state += std::to_string(note) + "-";
7  }
8  return state;
9  }
10
11 std::vector<int> MidiMarkovProcessor::markovStateToNotes(
12                  const std::string& notesStr)
13 {
14   std::vector<int> notes{};
15   if (notesStr == "0") return notes;
16   for (const std::string& note :
17          MarkovChain::tokenise(notesStr, '-')){
18     notes.push_back(std::stoi(note));
19   }
20   return notes;
21 }
```

Essentially, you create the state for a vector of notes by concatenating them with hyphens. Then, to get back to the vector from the state, tokenise on the hyphen. The code uses appropriate const and & syntax to make it clear the sent data is not edited and to ensure it is not unnecessarily copied.

25.2 Deciding when it is polyphonic: how long between notes?

Now, you can pass polyphonic states to the Markov model; you must work out when your plugin receives polyphonic MIDI in its MIDI buffer. You might think – well, I will simply check the timestamps of the notes, and if they are the same, it is a chord. If they are different, I will treat the notes as individual and pass them separately to the model. Unfortunately, the limitations of human beings might get in the way of that plan. Or at least human beings like me who cannot trigger notes in a sample-accurate manner. When people play a chord on a MIDI keyboard, the notes do not tend to start at precisely the same time, especially when you measure the start time in samples. They might not even arrive in the same call to processBlock. For example, my RME sound card will reliably run at a sample rate of 96kHz and a buffer size of 16 samples. If my DAW uses those settings, processBlock will only see 16 samples at a time, which is about 0.0002 seconds or 0.2ms. I wrote some test code to see how far apart in time the MIDI

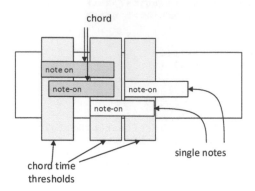

FIGURE 25.1

If notes start close enough in time, they are chords. If the start times fall outside a threshold, they are single notes. This allows for human playing where notes in chords do not happen all at the same time.

notes are when I play a chord, which you can put at the top of your processBlock function:

```
1 for (const auto metadata : midiMessages){
2     auto message = metadata.getMessage();
3     if (message.isNoteOn()){
4         DBG("note "
5                 << message.getNoteNumber()
6                 << " ts: "
7                 << message.getTimeStamp()
8                 << std::endl);
9     }
10 }
```

Running on a machine with a sample rate of 96kHz and buffer size of 256 I see the following output when I play a chord:

```
1 note 76 ts: 147
2 note 72 ts: 113
3 note 74 ts: 112
```

The time stamps are not the same – they are up to 30–40 samples or about 0.3ms apart at worst, and they potentially arrive in different calls to processBlock, so you need a way to remember notes across calls to processBlock. Another test is to play a fast monophonic sequence and to check the intervals between notes. That will tell you when you should treat things as monophonic.

25.3   Using the ChordDetector class

To solve this problem, I have implemented a class called ChordDetector, which implements some functionality to help with chord detection. Essentially, you tell it the threshold interval, which decides when notes are single or a chord. As discussed above, around 50ms of time works for me, but you may play faster or more accurately. Then you send notes to the ChordDetector, and if the notes are close enough in time, it accumulates them into a vector, which you can retrieve when you want to. You can find the ChordDetector class in code project 39.4.6. You can load that project and run tests on the ChordDetector class on the command line.

Place the ChordDetector.h and ChordDetector.cpp files into the src folder of your project. Then, you must edit your CMakeLists.txt file to include the new ChordDetector.cpp file in the build. Look for the target_sources command where you specify the .cpp files for the project. Add a line to target_sources(for the files:

```
target_sources(midi-markov-v1
    PRIVATE
    src/PluginEditor.cpp
    src/PluginProcessor.cpp
    ../MarkovModelCPP/src/MarkovManager.cpp
    ../MarkovModelCPP/src/MarkovChain.cpp
    src/ChordDetector.cpp
)
```

Once you have added it to your project, you can add a ChordDetector object to your PluginProcessor.h's private fields:

```
#include "ChordDetector.h"
// ... in the private section:
ChordDetector chordDetect;
```

Initialise chordDetect in your initialiser list in PluginProcessor.cpp since it needs a parameter sent to its constructor:

```
, chordDetect{0}
// constructor body here
{...
```

Then, set the max interval on chordDetect when prepareToPlay is called since that is when you can calculate the number of samples you want to allow for the chord threshold interval. In prepareToPlay:

```
double maxIntervalInSamples = sampleRate * 0.05; // 50ms
chordDetect = ChordDetector((unsigned long) maxIntervalInSamples);
```

To detect chords and send them to the Markov model appropriately formatted, replace the note-on section of your MIDI processing loop in analysePitches with this:

```
 1 if (message.isNoteOn()){
 2     chordDetect.addNote(
 3         message.getNoteNumber(),
 4         // add the offset within this buffer
 5         elapsedSamples + message.getTimeStamp()
 6     );
 7     if (chordDetect.hasChord()){
 8         std::string notes =
 9             MidiMarkovProcessor::notesToMarkovState(
10                 chordDetect.getChord()
11             );
12         DBG("Got notes from detector " << notes);
13         pitchModel.putEvent(notes);
14     }
15     ...
16 }
```

In that code, you call addNote to tell the chord detector about the note, query if it is ready to give you a chord with hasChord, and retrieve the chord with getChord. Then, you use notesToMarkovState to convert the notes vector into a suitable string for the Markov model. You might get a single note back from the chord detector if there is no chord – that is fine, as it should also be able to detect single notes. You can test this in Standalone mode. Click repeatedly on the on-screen keyboard to verify that you receive single notes. When you do that, you will see output something like this:

```
 1 ChordDetector::addNote 2205 n: 98 : 201216
 2 Got notes from detector 86-
 3 ChordDetector::addNote 2205 n: 88 : 223232
 4 Got notes from detector 98-
```

Now try holding the mouse button down whilst sliding up and down the on-screen keyboard. Let go occasionally, and you will see chords being detected, e.g.:

```
 1 Got notes from detector 76-77-79-81-
```

You should comment out the lines where you pull notes from the pitchModel when doing this test, as that code cannot cope with polyphonic states like this currently.

25.4   Generating polyphonic output

Now that you have fed polyphonic data to your Markov model, you can query the model and use the markovStateToNotes function to return the string states to vectors of note numbers. In your generateNotesFromModel function, update the code that generates notes – instead of this:

```
1  int note = std::stoi(pitchModel.getEvent(true));
```

Try this:

```
1  std::string notes = pitchModel.getEvent();
2  unsigned int duration = std::stoi(noteDurationModel.getEvent(true));
3  for (const int& note : markovStateToNotes(notes)){
4      juce::MidiMessage nOn = juce::MidiMessage::noteOn(1, note, 0.5f);
5      generatedMessages.addEvent(nOn, 0);
6      noteOffTimes[note] = elapsedSamples + duration;
7  }
```

Note that I used the same duration for all the notes in the chord. I set the velocity to 0.5f (We do not have a velocity model yet, but we will add one in the next section). If you try to pull a separate duration for each note, you will over-sample the model, causing the durations generated to not reflect the durations received by the model. You can try pulling multiple durations to see what happens.

25.5   Training a velocity model

We will add one more model to the improviser – a velocity model. This will model the sequence of velocities seen in the notes. It will make a big difference to the dynamics and expressiveness of the improviser's performance. You can probably figure out the steps here, but I will lead you through them anyway. If you want a clean, working polyphonic modeller project to start from, you can use example 39.4.7.

In PluginProcessor.h and cpp, add a model to the private section and initialise it in the initialiser list.

```
1  // in the header:
2  MarkovManager velocityModel;
3  // in the cpp above the constructor body:
4  velocityModel{}
```

Now add a new function to analyseVelocity:

```
1  // in the header:
2  void analyseVelocity(const juce::MidiBuffer& midiMessages);
3  // in the cpp:
4  void MidiMarkovProcessor::analyseVelocity(const juce::MidiBuffer&
       midiMessages)
5  {
6    // compute the IOI
7    for (const auto metadata : midiMessages){
8        auto message = metadata.getMessage();
9        if (message.isNoteOn()){
10           auto velocity = message.getVelocity();
```

```
11            velocityModel.putEvent(std::to_string(velocity));
12      }
13   }
14 }
```

Then in PluginProcessor.cpp, processBlock, call analyseVelocity:

```
1 analyseVelocity(midiMessages);
```

Then don't forget to call reset on the model in the resetModel function in PluginProcessor.cpp. That's it for the velocity modelling part.

25.6 Generating from the velocity model

To generate from the velocity model, query the model simultaneously as you query the duration model when generating notes. In PluginProcessor.cpp, generateNotesFromModel:

```
1 unsigned long duration = std::stoul(noteDurationModel.getEvent(true))
    ;
2 juce::uint8 velocity = std::stoi(velocityModel.getEvent(true));
3 ...
4 // then use it for the note:
5 juce::MidiMessage nOn = juce::MidiMessage::noteOn(1, note, velocity);
6 ...
```

That is it. In the repo guide, you can find the fully working model with polyphonic note modelling in example 39.4.8. I will now suggest experiments and extensions you can work on.

25.7 Experiment with the model

So now you have a polyphonic Markov model. Experiment with the plugin in a host environment. See if you can generate some interesting patterns. One way to train the model is to pass data from pre-existing MIDI files. You might already have a project in your DAW, or there are many MIDI files containing famous pieces of music, jazz solos, and such on the internet. If you have a DAW that can import MIDI files, you can load a MIDI file in, play it into your plugin, and have the plugin improvise along as it learns the file's contents.

25.8 Play-only mode

Once your model has learned some musical patterns, you might want to be able to generate from the model without it adding any more information to its state transition matrix (remember those?). This functionality should actually be relatively straightforward to implement. You need to switch on or off the bit of code in processBlock that calls putEvent on the Markov model. That is the part where the model learns.

You can achieve that by adding a 'learning on-off' toggle to your user interface and adding functions to your PluginProcessor to switch on or off calls to that putEvent function. I will leave you to work that one out yourself.

25.9 Loading and saving models

Another helpful feature is the ability to load and save models. If your model generates some really musical output, you might want to keep it for later. In fact, there is an expectation with plugins that the plugin will retain its state when you close and re-open a DAW project. You do not expect to have to re-program all your synths and effects every time you load a DAW project.

A basic implementation would allow the user to export models as files from the plugin and then load those files back in again. A better solution would automatically load and save the model by implementing the getStateInformation and setStateInformation functions, which are part of the plugin standard. If you implement those functions such that they save and restore the model to and from the sent data, the plugin will start with a trained model when you re-open a project in your DAW or other plugin host. You can call getModelAsString and setupModelFromString on MarkovManager to convert the models into strings. Then, you can concatenate the three models into a single string and convert it to the juce::MemoryBlock format needed for state saving. I will leave that for you to implement.

25.10 Progress check

At this point, you should have a polyphonic Markov model MIDI effect that you can load into a plugin host and train on MIDI input. The plugin will then generate polyphonic MIDI back out from its set of models, managing note pitch, velocity, duration and inter-onset intervals. I hope you were also able to carry out some experiments wherein you fed the model different types of training data. You may have even looked into more precise timing, loading and saving and other useful controls and functions for the plugin.

25.11 The end of this part of the book

Well done, you have reached the end of this part of the book. I hope you enjoyed working on the improviser and that you made some design decisions and customisations of your own along the way. As I mentioned earlier, I have been experimenting with these kinds of improvisers for many years. One version I built ran in the Supercollider environment, which you may have heard of. I set this improviser up to model patterns of MIDI controller data coming from an Akai MPD24 I was using. I was playing with drummer Paul Hession and Finn Peters, and I started the set manually, controlling all kinds of loopers and live effects feeding from mics on the musicians. Eventually, I switched my Markov improviser to generate mode and stepped away from the controls. It received more applause than I usually do, and it even made a musical joke at the end of our set, generating a ripple of laughter in the audience. Unfortunately, I have not been able to find the recording.

Part IV

Neural audio effects

26

Welcome to neural effects

Welcome to Neural Effects! In this part of the book, I will explain how you can use a neural network as an effects unit. Before presenting the full details of creating a neural audio effect later, I will provide some background information to help you understand the neural effects techniques. The background starts in this chapter, which presents a brief history of audio effects from the use of reverberant buildings through to neural networks emulating guitar amplifiers. Following this chapter, I will present further background material about what I call the 'DSP Trinity', namely finite and infinite impulse response filters and waveshapers. When we reach the coverage of neural effects, I will show you the full details of preparing training data, designing the network architecture and implementing the training in Python. Then, we will switch to C++ to develop the plugin. We will compare two methods for running neural networks: TorchScript and RTNeural. At the end of this part of the book, you will have a fully working neural network emulation of a guitar amplifier running in a plugin.

26.1   A brief history of audio effects

FIGURE 26.1
Tape manipulation was an early form of audio effect.

In 2020, Wilmering and colleagues presented a succinct history of audio effects[44]. They began by discussing reverberation as the first form of audio transformation, noting the long history of reverberant spaces, especially for religious worship. They then move from natural spaces with interesting acoustics on to artificial audio effects, organising the chronology of audio effects into four eras.

The first era covers audio transformation using sound recording and reproduction technology in the early 20th century. For example, Stravinsky manipulated phonogram tone arms to cause controllable distortion effects. Like Stravinsky before them, DJs still manipulate 'phonograms' today in the practice of turntablism. The era continued with the well-documented use of tape splicing techniques, for example Stockhausen's work. The practice of tape manipulation also continues today with artists such as Jamie Lidell[1] exploring creative interactions between tape and other music technology such as Eurorack.

The second era of audio effects concerns electro-mechanical effects. These are purpose-built hardware units which cause particular sound transformations. An example is the Leslie speaker from the 1930s, which featured a spinning high-frequency horn speaker. Fast-forwarding to the present day, electro-mechanical effects such as spring reverbs are still being produced.

The third era covers analogue signal processing, which uses analogue circuitry to transform the audio. This era, commencing in the 1930s, saw many new effects, such as frequency shifting, amplitude modulation, distortion, dynamic range compression and delay-line based reverberators. To connect this era to the present day, companies such as Warm Audio produce recreations of classic analogue effects, and companies such as Solid State Logic still make mixing consoles with analogue sound-transforming circuitry. Of course one must mention analogue synthesisers with their sound manipulating filters which are still mass-produced and very popular with musicians. Another connection between the analogue signal processor era and the present day is the range of digital emulations of analogue effects available as plugins. Emulating analogue processors using deep learning techniques is the most recent development here, and is of course the subject of this part of the book.

The fourth era identified by Wilmering et al. is the digital signal processing (DSP) era, which commenced for musicians and sound engineers at least, in the 1970s with the availability of rack-mountable digital effects. DSP occurs in the digital domain, where the signal is converted to a digital representation, and transformations are specified as digital processor instructions. The Eventide digital delay units were launched in 1971 and are early examples of digital effects. Guitarists gained access to cheaper digital delay line-based effects, such as choruses and flangers, in the 1980s via brands such as Boss. The 1990s saw the appearance of increasingly affordable and powerful DSP chips allowing for cheap multi-fx units such as those from Zoom. These were cheap enough that even I could afford one for my humble 1990s studio. I used a Zoom 1204 to process synth sounds from the synthesis chip on my Amiga 1200 computer.

The era of DSP took an interesting turn in 1996 when Steinberg launched version one of its Virtual Studio Technology (VST). This opened the door for developers to create all kinds of signal-processing plugins running on standard PC and Mac hardware, instead of requiring specialised outboard DSP systems. It ushered in the era of modern music production wherein producers could work

---

[1]http://jamielidellmusic.com/

completely 'in-the-box' in fully integrated digital audio workstation software environments such as Cubase.

An important theme in plugin software has always been emulating vintage, outboard effects. Increasingly accurate emulations of rare, valve-powered EQ and other units have been a beacon for the progress of musical digital signal processing techniques. An illustrative anecdote is that I was recently lucky enough to purchase an Apogee audio I/O box. It came with emulations of several vintage analogue effects and pre-amplifier circuits. It was only in a later software release that modern digital-style EQs and compressors were added to the plugins you could run on the device. The priority was clearly emulating analogue effects, and that from a company that made its name making digital anti-aliasing filters and analogue to digital converters.

Aside from developing digital studio production effects such as EQ and dynamic range compressors, another thread in the era of DSP has been the development of digital guitar amplifier emulations. These are typically hardware units which emulate one or more different guitar amplifiers using DSP techniques. The Line 6 Pod was an early unit launched in 1999. Later units like those from Kemper allow users to capture and emulate real amplifier setups. I will cover guitar amplifier emulation in more detail when we start working on a neural amplifier emulating effect.

26.2   The era of neural effects

Following Wilmering et al.'s fourth era of audio effects is the era of neural effects. Neural effects emerged in the early 2010s as a method to model signal processing systems using neural networks. At the same time, advances were being made in neural synthesis[7, 31], wherein neural networks *synthesize* audio signals, but our focus here is neural effects, wherein neural networks *process* audio signals. Much work on neural effects follows the trend of emulation discussed above and the earliest example of a neural network modelling an existing effect that I can find is Covert and Livingston modelling vacuum tube guitar amplifiers in 2013[9]. They stated that it demonstrated the idea's feasibility but that much more work was needed. Some of that work has now been done by people such as Alec Wright and Eero-Pekka Damskagg[46], working on guitar amplifier models in Finland around 2019 and Cristian Steinmetz and Martinez Ramirez's work in collaboration with Josh Reiss and others with reverbs and compressors at Queen Mary University in London[35, 40]. Damskagg was one of the people who worked on the Neural DSP system, which is one of the first commercially available hardware units to model guitar amplifiers using neural networks.

Much of this work relies upon established neural network architectures such as recurrent and convolutional models and you will learn all about these architectures in the upcoming chapters. But another stream of work in neural effects involves differentiable digital signal processing (DDSP), which allows more traditional signal processing components to be embedded within neural networks. The DSP components' parameters are then adjusted using neural network training methods. The idea is that the neural network gets a head start in learning an audio transformation as it starts with components which are designed to be able to carry out that kind of transformation, given the correct parameters. An example of this work came from Kuznetsov, Parker and Esqueda, who worked for Native Instruments[22]. They showed how to create efficient non-linear effects models for EQ and distortion using DDSP techniques.

With that, I will draw this short history of audio processing to a close. For further reading, the Wilmering et al. article is a good starting point for audio effects history, but there are also some fascinating articles online covering individual company and brand histories, such as Eventide and Boss. For neural effects, Vanhatalo et al. presented a survey in 2022 covering much of the work on neural guitar amp modelling, though we are going to implement one of those models for ourselves through the next few chapters[43]. If you are keen to read some more technical detail about state-of-the-art neural effects, you will find some very technical information about neural reverberators and some useful references in[24].

26.3 Progress check

This was a short chapter, but you should be able to differentiate between different ways of implementing audio effects and to place them in a chronological order.

27

Finite Impulse Responses, signals and systems

In this chapter, I will introduce some important digital signal processing (DSP) concepts. It is beyond the scope of this book to take a deep dive into the details of traditional digital audio effects design, but I am going to provide theoretical and code examples. For a detailed treatise on audio effects, I recommend Will Pirkle's book[33]. Here, I will cover the three main techniques used in digital audio effects and connect these to concepts found in neural network architectures. I refer to the three techniques as the 'DSP trinity': Finite Impulse Response Filters (FIR), Infinite Impulse Response Filters (IIR) and waveshapers. In the following sections, I will lead you through the basics of digital signal processing, trying to make it as accessible as possible by presenting the information in code, visualisations, and more traditional mathematical expressions.

27.1 Signals and systems

Digital signal processing revolves around the idea of signals and systems. We have a digital *signal*, e.g. an audio signal, and we pass it through some sort of digital *system*, which changes the signal somehow. Considering a single-channel audio signal, the signal x has a particular value at a certain point in time n: $x[n]$. The signal at the previous time step is $x[n-1]$. Now let's define a system. Imagine a system that represents a volume control. We can define the output of the system y at time n in terms of the input x and the volume level b:

$$y[n] = b.x[n] \qquad (27.1)$$

In case you prefer code, here is some code to compute output y based on input x and volume control coefficient b:

```
1 float b = 0.5; // volume level
2 float x[] = {0.1, -0.1, 0.2}; // signal
3 float y[3]; // output
4 for (int n=0;n<3;++n){
5     y[n] = b*x[n];
6 }
```

We can call that system a 'one-pole filter'. One-pole because the output calculation only considers a single input. Here is a two-pole filter which considers two inputs – the input now $x[n]$ and the input at the previous step $x[n-1]$. This time, instead of a single coefficient b we have two coefficients b_0 and b_1:

$$y[n] = b_0 x[n] + b_1 x[n-1] \qquad (27.2)$$

Then in C++ code:

```
int main(){
    float b0 = 0.5;
    float b1 = 0.5;
    float x[] = {0.1, -0.1, 0.2}; // signal
    float y[3]; // output
    for (int n=1;n<3;++n){
        y[n] = b0*x[n] + b1*x[n-1];
    }
}
```

This is a rudimentary low-pass filter because it averages two values, smoothing any large (fast, high frequency) changes. You can create all kinds of systems using as many poles as you want. The following expression is a general form for $y[n]$ for any number of coefficients. It might look a little intimidating with that $\sum$ sitting there, but really it is just a for-loop. The big $\sum$ means 'sum of' and i is the iterated value, and i goes from 0 up to N. N is the number of coefficients. So verbally, you might say the output at a point in time is the sum of previous inputs scaled by some coefficients.

$$y[n] = \sum_{i=0}^{N} b_i x[n-i] \qquad (27.3)$$

Some people prefer to unpack the equation like this, which is less elegant but maybe more readable:

$$y[n] = b_0 x[n] + b_1 x[n-1] + b_2 x[n-2] + ... + b_N x[n-N] \qquad (27.4)$$

How does the general form look in code? Here is a code version that will apply a system to a complete signal x. It is actually doing more than those equations since the equations only tell you $y[n]$ whereas the code computes the whole of y:

```
int main(){
    // number of coefficients
    int N = 3;
    // length of signals
    int sig_len = 6;
    // set of coefficients of length N
    float b[] = {0.25, 0.5, 0.25};
    // input signal of length sig_len
```

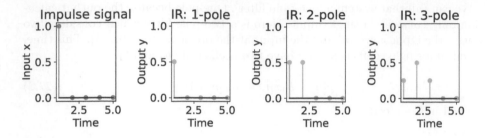

FIGURE 27.1

The impulse signal and the impulse responses of a one-pole, two-pole and three-pole system.

```
9       float x[] = {0.1,-0.1,0.2,0.5,0.25,0.1};
10      // zeroed out signal of length sig_len
11      float y[] = {0,0,0,0,0,0}; // output
12      // iterate over the signal, start N steps in
13      for (int n=N;n<sig_len;++n){
14          // iterate over the coefficients
15          // this is what the expression above describes
16          for (int i=0;i<N;++i){
17              y[n] += b[i]*x[n-i];
18          }
19      }
20  }
```

You can find that code in the repo guide section 39.5.3.

27.2 Impulse responses

The problem with digital signal processing is figuring out what the coefficients $b_{0...N}$ should be in order to create the effect you want. One way to do this is to take the effect you want to emulate and to find its *impulse response*. A system's impulse response is the system's output if you pass in a special type of 'test' signal called the impulse signal. The impulse signal is a single, maximum amplitude sample followed by silence. You can see it in the far left plot in figure 27.1. It sounds like a very short click. Figure 27.1 illustrates the impulse responses for the three systems we have seen so far. Why are impulse responses useful? The trick is that once you have the impulse response, assuming the system has certain characteristics, i.e., it is linear and time-invariant (more on that later), you can calculate the system's output in response to any signal you like.

Why can you calculate the output of a system for any signal, if you have the impulse response? The answer is that there is a direct relationship between the coefficients of a system (b_0, b_1 etc.) and the impulse response. In fact, impulse responses *are* the coefficients, at least for the types of systems we have seen so far. Think back to the 2-pole averaging filter we discussed earlier. In the equation and code we defined the coefficients as 0.5 and 0.5:

$$y[n] = 0.5x[n] + 0.5x[n-1] \qquad (27.5)$$

If you put the impulse signal through that, you get $[0.5, 0.5]$ out, as shown in figure 27.1 – the impulse response is the same as the coefficients. So if we can capture an impulse response of a system by sending the impulse signal in and recording the output, we can use the output as coefficients and compute the effect of that system on any signal. When I first encountered this concept and implemented it in code, I was really blown away! Why? because you can capture your own impulse responses for effects by passing a click through them and recording the output. For example, you can play a click in a room and record the response. Then you can apply that response to any signal and it will sound as it would if you played the signal in the same room. We will do exactly that later in the chapter when we discuss convolution, which is the name of this process of applying impulse responses to systems.

27.3   Linear time invariant

Before we dive in and experiment with convolution, I would like to clarify what LTI means. Linearity and time-invariance are important properties in DSP systems because the process of convolution, wherein you can apply a system to any signal given its impulse response, only works for LTI systems. I am going into these concepts here because they explain why it is difficult to model valve guitar amplifiers and distortion pedals, which we will be doing later. Linearity means that if you double the input values, the output values will also be doubled. It also applies to any other scalar: double, triple, 0.2555 etc. Here is an example of linearity and scaling when the system is a simple 0.5 multiplier, i.e. has impulse response $[0.5, 0, 0, 0]$:

$$y[n] = b_0 x[n] \qquad \text{the system}$$
$$b_0 = 0.5 \qquad \text{the coefficient}$$
$$0.5 \to 0.25 \qquad \text{input } 0.5$$
$$0.6 \to 0.3 \qquad \text{larger input } 0.6 \tag{27.6}$$

Consider a real-world example, e.g. a valve guitar amplifier. As you increase the gain, which is equivalent to scaling the input in a certain sense, the output signal changes. But – it does not just get louder – it also gets more distorted. So the valve guitar amplifier is not linear. Guitarists like non-linear amplifiers because increasing the gain varies the tone in musical ways.

Now think of a professional speaker amplifier that you might use in a recording studio. You do not want it to change the character of the sound based on the gain, because that would make it difficult to produce a correct mix – the mix would change based on the gain. So linearity is a desirable property in such an amplifier. We might say that a professional studio amplifier is more linear than a valve guitar amp. With digital systems, it is possible to achieve perfect linearity, in the digital part at least.

Another part of linearity is additivity: if you pass two signals through the system on different occasions, then capture and sum the outputs, you will have the same output as if you passed in the sum of the two signals as an input. Here is a simple demonstration of additivity, where the important thing is the fact that $0.375 = 0.25 + 0.125$:

$$y[n] = b_0 x[n] \qquad \text{the system}$$
$$b_0 = 0.5 \qquad \text{the coefficient}$$
$$0.5 \to 0.25 \qquad \text{input } 0.5$$
$$0.25 \to 0.125 \qquad \text{input } 0.25$$
$$0.75 \to 0.375 \qquad \text{input } 0.5 + 0.25 \tag{27.7}$$

What about time invariance? This means that passing a signal in with a delay does not change the output signal in any way aside from delaying it. The behaviour of the system does not change over time. So the 1930s Leslie speaker we discussed earlier which has a spinning treble horn is not time-invariant. Here is an example of time invariance where two signals x_1 and x_2, with x_2 the same as x_1 but with a single sample delay are passed through the simple system $y[n] = b_0 x[n]$:

$$y[n] = b_0 x[n] \qquad \text{the system}$$
$$b_0 = 0.5 \qquad \text{the coefficient}$$
$$x_1 = [0.25, 0.5, 0.0] \qquad \text{input 1}$$
$$x_2 = [0.0, 0.25, 0.5] \qquad \text{input 2 (delayed)}$$
$$y_1 = [0.125, 0.25, 0.0] \qquad \text{output 1}$$
$$y_2 = [0.0, 0.125, 0.25] \qquad \text{output 2 (delayed)} \qquad (27.8)$$

But what about parameterisation? If a digital signal processing system has a parameter that changes its behaviour, is it an LTI system? Think of a reverb effect which has a room size parameter. As you adjust the parameter, the impulse response of the system changes. The concept still holds – you just need to consider the reverb as being several different linear, time-invariant systems. As you adjust the room size parameter, you are switching to a different system with different coefficients in its impulse response. If you do not adjust the parameter, the system is linear and time-invariant. If you adjust the parameter, you are switching to a different system.

27.4   Progress check

So far, this section has been quite theoretical. But you can do some experiments with the code snippets you have seen. Can you verify that the three-pole filter system is linear and time-invariant, for example? If you are feeling adventurous you could implement a simple plugin in JUCE which allows you to experiment with filters with different numbers of poles. You could create a user interface which allows the user to specify a filter by drawing an impulse response.

28

Convolution

In this chapter, I am going to discuss convolution. Convolution is the process that allows you to compute the effect of passing a given signal through a given system. We will look at two ways of carrying out convolution – in the time and frequency domains. The time-domain process maps closely to the equations we have seen already, making it easy to understand. The frequency domain version is far more efficient but it is a less direct interpretation of the equations we saw in the previous chapter. As you work through the chapter you will see various code examples, initially running on the command line to process audio files with convolution. At the end of the chapter, you will see how you can implement efficient convolution using the built-in DSP components available in the JUCE API, resulting in a convolution reverb plugin.

28.1 System setup

Before we start on the work in this chapter, you will need to set up your C++ environment including CMake and JUCE. Please refer to the instructions in Part I. From this point, I will assume you are ready to build and run C++ programs.

28.2 Loading signals and systems

Let's start by writing a C++ program to load an audio file into memory for processing. We will use the tinywav library by Martin Roth[1]. As a brief digression, I worked with Martin Roth in the noughties on an automated sound synthesiser programmer called Synthbot[47]. This happened after we met at an evolutionary computing conference in Valencia, Spain and got into some great discussions about

---

[1] https://github.com/mhroth/tinywav

the possibilities for audio processing and evolutionary algorithms. Since then, I lost touch with Martin, but when preparing this chapter, I needed a simple way to load audio files into a C++ program with minimal dependencies. The WAV format is perhaps the most common non-compressed audio file format, but it can be a bit tricky to work with because it has several ways to format its data. I found the tinywav library on Github by chance and was delighted to see who had written it – my old collaborator Martin Roth. Tinywav does precisely what I wanted for these convolution examples – it loads a WAV file into a list of floats. It also provides functions to save floats back out to WAV format.

Right, back to the main task. Here is a program that uses tinywav to load a WAV file into memory:

```
1 #include "../../tinywav/myk_tiny.h"
2 #include <string>
3 #include <vector>
4 #include <iostream>
5
6 int main(){
7     std::vector<float> audio = myk_tiny::loadWav("audio/drums_16bit.
      wav");
8 }
```

This code is described in the repository section 39.5.1. The myk_tiny namespace also contains a function saveWav:

```
1 static void myk_tiny::saveWav(
2     std::vector<float>& buffer,
3     const int channels,
4     const int sampleRate,
5     const std::string& filename);
```

Go ahead and build the project. See if you can load and save some audio files. Verify the files are saved correctly by loading them into an audio editor such as Audacity. I have only tested it with mono files with a sample rate of 44,100Hz, as those are the files I am interested in for now. Note that the saveWav function uses the TW_INT16 format, two bytes signed integer. Two-byte signed integers fall in the range of −32768 to 32768. The loadWav function does not specify the format, so it converts the samples into the default format for tinywav, a 32-bit/four-byte IEEE float. The float values will be from −1 to 1 once they appear in your program. You can verify this by scanning through the vector and checking the lowest and highest values. Do some experiments to confirm that things are loaded and saved correctly and that the various number formats yield values in the ranges you expect.

28.3   Simple time-domain convolution

Now that you can load and save audio files, you are ready to try some convolutions. Here is a time domain convolution function which applies the impulse response, received as parameter 'bs' to each sample in the input signal (xs), summing the result as it goes:

```
1  std::vector<float> conv(std::vector<float> xs, std::vector<float> bs)
      {
2      // prepare a vector to store output y
3      std::vector<float> y(xs.size() + bs.size(), 0.0f);
4      int b_count = bs.size();
5      // iterate over signal x
6      for (int n=0;n<xs.size();++n){
7          if (n % 10000 == 0){
8              printf("%i of %lu \n", n, xs.size());
9          }
10         y[n] = 0;
11         // iterate over systems' coeffs b
12         for (int b=0;b<n && b < b_count;++b){
13             y[n] += bs[b] * xs[n-b];
14         }
15     }
16     return y;
17 }
```

This is a quite mechanical implementation of the 'apply weights and sum' equation shown below:

$$y[n] = \sum_{i=0}^{N} b_i x[n - i] \tag{28.1}$$

For clarity, in that equation, x is the input signal, y is the output signal, n is the current time step, N is the number of coefficients, i is the current coefficient index and $b_{(}0...N)$ is the set of coefficients. If you look carefully, you will notice that the code does more than that equation as the code also iterates over the whole of x, so to fully represent the code we need another $\sum$:

$$y = \sum_{i=0}^{N} \sum_{n=0}^{M} b_i x[n - i] \tag{28.2}$$

Now we are iterating over the input signal from position 0 to position M, which is the length, or $xs.size()]$ on line 6 of the code above.

Experiment with the code. First, try creating the impulse response vector 'b' with some simple coefficients, e.g.:

```
1  void experiment1(){
2      std::string xfile = "../audio/drums_16bit.wav";
3      std::string yfile = "../audio/drums_expt1.wav";
4      std::vector<float> x = myk_tiny::loadWav(xfile);
5      // simple moving average low pass filter
6      std::vector<float> b = {0.5, 0.5, 0.5};
7      std::vector<float> y = conv(x, b);
8      myk_tiny::saveWav(y, 1, 44100, yfile);
9  }
```

You can find example code in the repo guide section 39.5.3. With my example drum sample 'drums_16bit', which you can find in the audio folder for this part of the book, I find that the resulting signal is somewhat clipped – after the convolution, some of the values are > 1. When you write these to a file, it 'ceils' them to 1. To fix the clipping, I divided all values by the sum of the coefficients. This is the maximum achievable value for a sample in the output signal. Here is a simple amp function to apply an in-place scale to a vector of floats:

```
1  void amp(std::vector<float>& xs, float amp){
2      for (int n=0;n<xs.size();++n){// iterate over signal x
3          xs[n] = xs[n] * amp;
4      }
5  }
6  ... then in main ...
7  std::vector<float> y = conv(x, b);
8  amp(y, 1/1.5);// scale by reciprocal of sum of 0.5, 0.5, 0.5
9  myk_tiny::saveWav(y, 1, 44100, yfile);
```

You could actually go a step further and implement a function to sum the coefficients:

```
1  float sumCoeffs(std::vector<float>& coeffs){
2      float sum = 0.0;
3      for (float& f : coeffs) sum += f;
4      return sum;
5  }
```

Swap it out for the hard-coded value 1.5 and it will work for any set of coefficients:

```
1  ...
2  // instead of
3  // amp(y, 1/1.5);
4  // do this:
5  amp(y, 1/sumCoeffs(b));
6  ...
```

Earlier, I mentioned that this moving average filter causes a low pass effect – it removes some of the high frequencies. We can verify that by looking at the spectrum of the sound before and after. You can use Audacity to plot a spectrum – use plot spectrum from the analyse menu. I used a Python script, which you can find in the repository section 39.5.4. The result can be seen in figure 28.1. Note

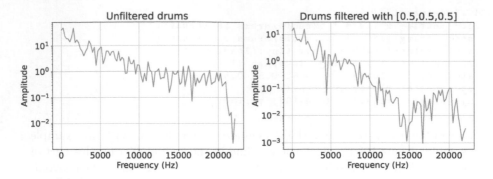

FIGURE 28.1
Original drum loop spectrum on the left, filtered version on the right. High frequencies have been attenuated in the filtered spectrum.

how the image on the right shows a dent starting at around 15,000Hz caused by the moving average filter. Experiment with different coefficients and see what variation you get in the filtered spectrum. Try adding more coefficients, negative coefficients and so on. Can you create an echo effect somehow? The take-home message is that you can make different filters and delay effects using various combinations of coefficients.

It is beyond the scope of this book to fully explore all the possibilities convolution can offer, but we will continue a little further. Firstly, we will see a faster implementation of convolution based around the FFT. Then we will see a real-time, JUCE plugin implementation of convolution based on DSP components available in the JUCE library.

28.4 Time-domain convolution is slow

Before moving on from time-domain convolution, I would like you to investigate how long it takes to carry out time-domain convolution. I have created a simple project to test the performance with different numbers of coefficients, which you can find in the repo guide section 39.5.5. It has some somewhat hacky code to generate increasingly large impulse responses and then to time how long it takes to apply those responses to a signal loaded from a WAV file. Here is the program code – starting from 1000 coefficients, it increases to 200,000. With a sample rate of 44,100Hz, 200,000 samples represent about 4.5s of audio. If you think about how long a reverb might last, e.g. if you clap you hands in a cave, 4.5s is not an unreasonable impulse response.

```
std::string xfile = "../../audio/drums_16bit.wav";
std::string yfile = "../../audio/drums_expt1.wav";
std::vector<float> x = myk_tiny::loadWav(xfile);
// how many seconds in the file
float file_len_seconds = x.size() / 44100.0f;
// simple moving average low pass filter
for (float i=1000;i<200000;i+=1000){
    std::vector<float> b{};
    for (int j=0;j<i;++j){
        b.push_back(0.5);
    }
    auto start = std::chrono::high_resolution_clock::now();
    std::vector<float> y = conv(x, b);
    auto stop = std::chrono::high_resolution_clock::now();
    auto duration =
        std::chrono::duration_cast<std::chrono::milliseconds>(stop -
        start);
    // duration is how long it took to process the entire audio file.
    float dur_per_second = (duration.count()/1000.0f)  /
    file_len_seconds;
    float dur_per_coefficient = dur_per_second / b.size();
    std::cout << "With IR len " << b.size()
            << " conved " << file_len_seconds
            << "s file in " << (duration.count() / 1000.0f)
            << "s" << std::endl;
    }
```

I see output like this on my 10th gen Intel CPU Linux machine, using a 'Release' version of the build, processing a 2.8s drum loop with varying length impulse responses:

```
With IR len 1000 conved 2.82782s file in 0.142s
With IR len 2000 conved 2.82782s file in 0.234s
With IR len 3000 conved 2.82782s file in 0.362s
With IR len 4000 conved 2.82782s file in 0.46s
...
With IR len 25000 conved 2.82782s file in 2.675s
With IR len 26000 conved 2.82782s file in 2.885s
With IR len 27000 conved 2.82782s file in 2.889s
With IR len 28000 conved 2.82782s file in 2.929s
```

This tells me that I can potentially run up to 25,000 coefficients in real time, as it takes less time than the length of the file to process 25,000 coefficients. A quick test on my basic 2020-model M1 Mac Mini showed that it can run about 27,000 coefficients (an impulse response of about 0.5s) in real-time. Not bad! But a typical reverb effect can reverberate for much longer than 0.5s, so how to achieve greater performance?

28.5 Frequency domain convolution

The answer to improved convolution performance is to convert the signal and the system to a frequency domain representation, to convolve, and then to revert the result to the time domain. This is faster because the convolution operation becomes a simple matrix multiplication in the frequency domain instead of the many array scaling and summing steps needed in the time domain. It is beyond the scope of this book to explain exactly why these two are equivalent, but I will show you an example program that convolves in the frequency domain and compare its speed.

The repo guide describes the frequency domain convolution program in section 39.5.6. It uses another lightweight, low-dependency library to carry out the time-to-frequency domain conversion: Simple-FFT by Dmitry Ivanov[2]. I have not met Dmitry, unlike Martin Roth but the library is certainly simple and fast. There are some caveats associated with frequency-domain convolution, though. Firstly, the program convolves the whole signal in one go, so the impulse response and the signal must be the same length to allow the matrix multiplication. For simplicity, my program just zero-pads the shorter of the two vectors (signal or system), to make them the same length. This involves adding zeroes to the end of the shorter vector. The second caveat is that the FFT algorithm requires the signal's length to be a power of two. For simplicity, I just truncated the longer of the signal or system to the nearest power of two below its length. The end result of the truncation and zero padding is a system and a signal vector of the same size, where the length is a power of two.

How fast does this alternative convolution algorithm convolve? If I convolve the 124,707 sample/ 2.82s long drum loop with an impulse response, the nearest power of two to the length of the signal is 65,536. Remember that the time-domain version ran fast enough to do 25,000 coefficients in real time. The complete time required to resize and zero-pad the signal and system, to convert them to the frequency domain using the FFT, to convolve, and then to inverse FFT the result back to a time domain signal is as follows:

```
1  With IR len 65536 conved 1.48608s file in 0.012s
```

So 12ms to convolve 65,536 samples of audio (1.48 seconds) with 65,536 coefficients. My Apple M1 Mac Mini completes in 7ms. The time domain version takes 2.012s or 2,012ms to compute a 65,535 convolution on a 65,5536 input, or about 168 times as long. This is a remarkable performance boost. Do some experiments to verify that the frequency domain convolution is, in fact, doing the same thing

[2]https://github.com/d1vanov/Simple-FFT

as the time domain convolution. Try convolving signals with both techniques and comparing the results by listening to them.

28.6 Convolution in JUCE

Now that you have gained some familiarity with convolution, I will show you how to implement convolution in a plugin using some ready-made DSP components from the JUCE API. Convolution is well known for its use in reverb plugins; the first time I encountered a convolution reverb was using a Waves plugin in around 1998. You can hear a reversed convolution reverb effect in my track 'E-20 Crunch' on my SuperUser album released in 2000 on Rephlex Records. The reverb plugin was quite expensive then – I think it cost about three times as much as the second hand Amiga computer I used for sequencing and sound synthesis at the time. I actually used a friend's computer to create the effect. It is amazing that now it is possible to recreate a basic version with about 20 minutes of C++ hacking and the JUCE framework. Thanks, Jules!

28.6.1 Prepare the plugin project

You can begin the work in this section with the plugin starter project described in the repo guide section 39.2.2. Load up the CMakeLists.txt file and edit the target name, which will start as 'minimal_plugin'. You should call it something with the word convolution in it. Replace all instances of that target name in the CMakeLists.txt file. Also, set your PLUGIN_CODE to something unique.

Next, in the juce_add_plugin command, set 'IS_SYNTH' to 'FALSE'. This will allow the plugin to be an effects unit. Edit the target_link_libraries block by adding the 'juce::juce_dsp' module. My target_link_libraries block looks like this:

```
target_link_libraries(convolution_plugin
    PRIVATE
        juce::juce_audio_utils
        juce::juce_dsp
    PUBLIC
        juce::juce_recommended_config_flags
        juce::juce_recommended_lto_flags
        juce::juce_recommended_warning_flags)
```

At this point, you should attempt to build the plugin. Depending on which IDE you are using, take the appropriate steps to generate and build the IDE project.

28.6.2 Create the convolution DSP chain

In your project, open up the PluginProcessor.h file and add the following lines to the private section of the Plugin class you find there:

```
1 juce::dsp::ProcessorChain<juce::dsp::Convolution> processorChain;
```

You can think of a ProcessorChain as a series of effects units through which you can send signals. In this case, the series will only have one unit: our convolution effect. Next, open up PluginProcess.cpp and add the following line to the constructor:

```
1 auto& convolution = processorChain.template get<0>();
```

This gives you access to the actual processor from the effects chain so that you can interact with it. You need to interact with it so you can tell it to load an impulse response:

```
1 juce::File impFile{"/fullpath/to/an/impulseresponse/here"};
2 convolution.loadImpulseResponse (impFile,
3                                  juce::dsp::Convolution::Stereo::yes,
4                                  juce::dsp::Convolution::Trim::no,
5                                  0);
```

Make sure you set impFile to a path that exists. For now, we will hard-code it. The following line calls loadImpulseResponse. The final option, '0', tells it to keep the complete impulse response. If you do not have an impulse response file, you can create one by playing a loud click from your speaker and recording the result. Impulse responses are also available from online sources such as York University's excellent OpenAIR repository[3].

28.6.3 Configure the chain for audio processing

Now you have created the chain and loaded an impulse response, you need to set it up for audio processing. First, you need to implement the reset function on your plugin. The base plugin comes with an implementation of reset, but you can override it. In PluginProcessor.h:

```
1 void reset() override;
```

Then the implementation in PluginProcessor.cpp:

```
1 void TestPluginAudioProcessor::reset()
2 {
3     processorChain.reset();
4 }
```

This just passes on the reset message to the convolution processor. The final

---

[3]https://openairlib.net/

things to implement are prepareToPlay and processBlock. prepareToPlay is the function called by the plugin host when it wants to tell the plugin what the audio configuration is. processBlock is the function called by the host every time it wants the plugin to process a block of audio. In PluginProcessor.h, you should already have the declaration of these functions, so you can go straight to PluginProcessor.cpp and add their implementations. Firstly, prepareToPlay:

```
1  // put this in prepareToPlay:
2  //How many channels?
3  const auto channels = jmax (getTotalNumInputChannels(),
4                              getTotalNumOutputChannels());
5
6  //Tell the processor about the audio setup
7  processorChain.prepare ({ sampleRate,
8                            (uint32) samplesPerBlock,
9                            (uint32) channels });
10 reset();
```

That code informs the convolution processor about the sample rate, block size and channel count. Without digging into the JUCE convolution implementation, I assume this will set up the FFT block size and such. Next, put this in processBlock in PluginProcessor.cpp, removing anything else that is there:

```
1  ScopedNoDenormals noDenormals;
2  //How many channels?
3  const auto numChannels = jmax (getTotalNumInputChannels(),
4                                 getTotalNumOutputChannels());
5
6  //Convert the buffer we were sent by the plugin host (buffer) into
7  // an AudioBlock object that the convolution processor can understand
   :
8  auto inoutBlock = dsp::AudioBlock<float>
9                    (buffer).getSubsetChannelBlock (0, (size_t)
   numChannels);
10
11 // carry out the convolution 'in place.'
12 processorChain.process (
13         dsp::ProcessContextReplacing<float> (inoutBlock));
```

Now compile, debug your errors and run it in your plugin host. This setup is most useful for long impulses like reverberations. Once everything is running you can try out some extensions and experiments. For a start, the GUI is essentially non-existent. Can you add some controls? For example, a dry/wet mix is a common control to find on a reverb unit. Can you add an option to load an impulse response chosen by the user? Then how about controls to adjust the impulse response? For example, you could have a 'length' control that somehow generates permutations of the impulse response that are shorter or longer. Think about how you might make a shorter version of a given impulse response – you could fade it out and chop off the remainder, for example, or you could create a new version by skipping every other sample. How about a longer version? Over to you!

28.7 Progress check

In this chapter, you should have worked through some command-line programs that carry out convolution on WAV files. You should be able to explain the difference between time domain and frequency domain convolution and say which one is faster. You should also know how to use classes from the JUCE API to implement a convolution plugin. I hope that you also customised your convolution plugin with some extra controls.

29

Infinite Impulse Response filters

In this chapter, you will find out about infinite impulse responses and how they differ from the finite impulse responses seen in the previous two chapters. We will start with a low-level C++ implementation and then move towards a real-time plugin implementation using JUCE. You will learn about filter coefficients and frequency responses. You will see how you can use JUCE's DSP module functionality to implement IIR filters efficiently with preset coefficients for different types of filters.

29.1   Infinite Impulse Responses (IIR)

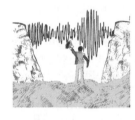

FIGURE 29.1
Infinite impulse responses are powerful!

In the previous chapter, you learned about convolution, which involves applying an impulse response to a signal. Hopefully, you remember that the output is computed based on a weighted sum of previous inputs to the system. If the input fades to silence (as the impulse signal does after its first sample), the output will eventually become silent. Therefore, all the impulse responses we have seen so far are *finite* impulse responses, or FIRs.

There is another type of impulse response, which, unlike the FIR, is not guaranteed to fade to zero. This is the Infinite Impulse Response (IIR). The trick with IIRs is that the output is computed based on the previous inputs *and* the previous outputs – the previous outputs are fed back into the input – they exploit the power of *feedback*. There is a fascinating cross-disciplinary history of feedback systems in art, science and engineering. I recommend cybernetics as an interesting launching-off point. Pickering's 2002 paper provides some insight into the activities of British cybernetics pioneers Ashby, Pask and Beer in the mid-20th century, who were extremely interested in feedback[32]. In a related parallel universe, Jamaican producers such as King Tubby and Augustus Pablo pioneered feedback

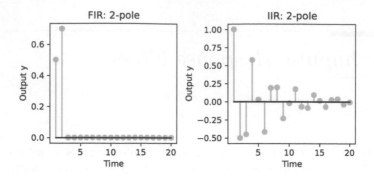

FIGURE 29.2

Pole for pole, IIR filters generate much richer impulse responses. The left panel shows a two-pole, FIR. The right pane shows a two-pole IIR.

effects in music production during the 1970s. Yoganathan and Chapman wrote an interesting article relating this music production practice to electroacoustic music practice occurring at a similar point in time but in a very different context[50].

Returning to the topic of digital signal processing, the thing to know about IIR and feedback is that compared to FIR and convolution, it allows you to create more drastic filtering and delay effects with far fewer coefficients. Figure 29.2 illustrates this, showing finite and infinite two-pole impulse responses, where the infinite two-pole is far more complex. This means IIR filters are more efficient to run.

As an example, in the early days of DSP-based effects for music production, it was impossible to convolve complex, finite impulse responses such as room reverb responses with signals in real time. Part of this was computational complexity and part was the cost of the memory needed to store the impulse responses.

Instead of FIR filters, early real-time digital reverbs such as 1979's famous Model 224 from Lexicon, used IIR techniques extensively. Convolution and FIR were only available for non-real-time use. It was not until 1990 that DSP hardware was sufficiently fast and RAM sufficiently cheap to allow for convolutional reverbs. The Sony DRE S777 was a £20,000, 2U rack-mount unit which could load impulse responses from a CD-ROM and which Sound on Sound magazine described as 'Bulky, heavy and hot' but with 'Stunningly believable room signatures'[1].

IIR filters are also used extensively in digital sound synthesisers where filters with musical characteristics such as strong resonance peaks are desirable, as IIR filters are excellent for these kind of applications. This is especially true when

---

[1]https://www.soundonsound.com/reviews/sony-dre-s777

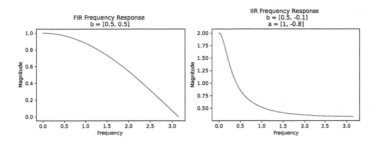

FIGURE 29.3
Comparison of two pole FIR filter (left) and IIR filter (right). The IIR filter has a more drastic response.

modelling classical analogue synthesiser filters, which are often based on feedback circuits themselves. Figure 29.3 aptly illustrates the power of IIR filters compared to FIR filters with a similar number of poles. You can see that the frequency response of a simple two-pole IIR filter is far more drastic than that of a two-pole FIR filter.

Music technologists should be familiar with and possibly live in fear of feedback and its power. You only need to work with music technology for a short time before you inflict painful feedback upon yourself or others. I personally ripped a pair of headphones off my head just last week after a granular looping sampler effect combined badly with a reverb effect in a SuperCollider patch I was working on.

29.2 One-pole IIR filter in C++

Expressing feedback mathematically is less painful (physically, at least), relatively straightforward, and similar to the expressions for finite impulse responses you saw earlier. A simple, one-pole infinite impulse response can be expressed as a difference equation as follows:

$$y[n] = a.y[n-1] + b.x[n] \tag{29.1}$$

Note how we used an a and a b to differentiate between coefficients acting on input x and those acting on output y. I have seen different versions of this equation. The b is sometimes written first with the a term subtracted. As a computer scientist, I do not care, as I can use the same code and just make the coefficients negative, depending on which book I am reading!

In code, that filter looks like this:

```
1  int main(){
2      float a = -0.9;
3      float b = 0.5;
4      float x[] = {0.1, -0.1, 0.2, 0.5, 0.25, 0}; // signal
5      float y[] = {0, 0, 0, 0, 0, 0};
6      for (int n=1;n<6;++n){
7          y[n] = a*y[n-1] + b*x[n];
8          printf("x[%i]=%f y[%i]=%f\n", n, x[n], n, y[n]);
9      }
10 }
```

Note that the a coefficient acts on previous output $y[n-1]$ and the b coefficient acts on current input $x[n]$.

On my system, I see the following output:

```
1  x[1]=-0.100000 y[1]=-0.050000
2  x[2]=0.200000 y[2]=0.145000
3  x[3]=0.500000 y[3]=0.119500
4  x[4]=0.250000 y[4]=0.017450
5  x[5]=0.000000 y[5]=-0.015705
```

Now, let's adapt the code to continue computing the output array y for longer to see the effect of the 0.9 feedback coefficient. In the following code, I computed 100 values of y, using zero padding for input x:

```
1  int main(){
2      float a = -0.9;
3      float b = 0.5; // feedback
4      float x[] = {0.1, -0.1, 0.2, 0.5, 0.25, 0}; // signal
5      float y[100]; // output - we'll run it for 100 samples
6      y[0] = 0;
7      for (int n=1;n<100;++n){
8          float xn = 0; // zero pad
9          if (n < 6) xn = x[n]; // unless there x has a value for n
10         y[n] = a*y[n-1] + b*xn;
11         printf("x[%i]=%f y[%i]=%f\n", n, xn, n, y[n]);
12     }
13 }
```

To avoid creating a large, zero-padded input array x, I just decided if I should obtain x_n from x or set it to zero.

```
1  x[1]=-0.100000 y[1]=-0.050000
2  x[2]=0.200000 y[2]=0.145000
3  x[3]=0.500000 y[3]=0.119500
4  ....
5  x[26]=0.000000 y[26]=0.001718
6  x[27]=0.000000 y[27]=-0.001547
7  ....
8  x[97]=0.000000 y[97]=-0.000001
9  x[98]=0.000000 y[98]=0.000001
10 x[99]=0.000000 y[99]=-0.000001
```

Even with one pole and only six non-zero inputs, you can see that the signal continues to register after 100 steps. The working code for this program is described in the repo guide section 39.5.8. Experiment with the code – what happens with increasing values for coefficient a, for example?

29.3   Two-pole IIR filter in C++

Now, we will add another pole. Here is a two-pole version of the expression, which considers two previous outputs $y[n-1]$ and $y[n-2]$ along with one previous input $x[n-1]$ and the current input $x[n]$:

$$y[n] = a_1 y[n-1] + a_2 y[n-2] + b_1 x[n] + b_2 x[n-1] - \qquad (29.2)$$

Again, I have written it as a simple addition, though you might see it written with the b components as a subtraction. Here is some code to compute the output of that two-pole infinite impulse response filter with some specific coefficients:

```
int main(){
    float a1 = 1; // output coefficient 1
    float a2 = -0.8; // output coefficient 2
    float b1 = 0.5; // input coefficient 1
    float b2 = -0.1; // input coefficient 2
    float x[] = {0.1, -0.1, 0.2}; // signal
    float y[3]; // output
    for (int n=2;n<3;++n){
        y[n] = a1*y[n-1] + a2*y[n-2] +
                b1*x[n] + b2*x[n-2];
    }
}
```

You can experiment with this code yourself, for example, computing a longer output signal. You could attempt to integrate the code with the tinywav example code described in the code repo section 39.5.1 so you can process an entire audio file with your IIR filter.

29.4   General form for IIR and FIR filters

Here is the general difference equation for an IIR filter, allowing for any number of poles. I have put it in the 'subtract' form seen on Wikipedia and in certain DSP textbooks:

$$y[n] = b_0x[n] + b_1x[n-1] + b_2x[n-2] + \ldots + b_Mx[n-M]$$
$$- a_1y[n-1] - a_2y[n-2] - \ldots - a_Ny[n-N] \qquad (29.3)$$

This equation looks beastly, but it expands from the one- and two-pole examples above. Sometimes, a normalisation component is added, which is the reciprocal of the sum of all the a and b coefficients. This is similar to how we normalised the convolution before. Here is an implementation of the general IIR filter, which you can find in the repo guide section 39.5.9

```
1  std::vector<float> x = myk_tiny::loadWav(
2                  "../../audio/drums_16bit.wav");
3  std::vector<float> as = {0.5, 0.1, 0.2};
4  std::vector<float> bs = {0.1, -0.7, 0.9};
5  std::vector<float> y(x.size(), 0.0f);
6  for (auto n=as.size();n<x.size(); ++n){
7      float yn = 0;
8      // weighted sum of previous inputs
9      for (auto bn=0;bn<bs.size();++bn){
10         yn += bs[bn] * x[n - bn]; // acting on input x
11     }
12     // weighted sum of previous outputs
13     for (auto an=0;an<as.size();++an){
14         yn -= as[an] * y[n - an]; // acting on output y
15     }
16     y[n] = yn;
17 }
18 myk_tiny::saveWav(y, 1, 44100, "../../audio/iir_test.wav");
```

29.5 Frequency responses

Looking at the various code examples in the previous sections, you may wonder how you know what the coefficients should be. This was easy for the FIR filters as the coefficients were exactly the impulse response. For FIR filters, if you can obtain the impulse response of a system, then you can convolve any signal with that system. This is not true for IIR systems. There is not such a direct way to obtain the coefficients. The general process of identifying IIR filter coefficients is referred to as 'Filter Design'. A full exploration of classical filter design techniques is beyond the scope of this book, but to give you some idea of the process, you usually start with a frequency response, which is a numerical description of how the filter you desire affects different frequencies in the signal. E.g. does it remove low frequencies? How much? What is the slope?

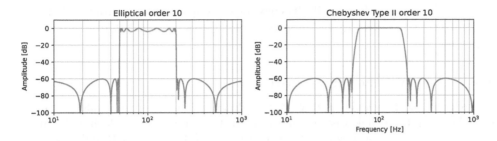

FIGURE 29.4
Two types of IIR filter and their frequency responses. IIR filter design is a compromise.

For example, say you want a band-pass filter. Band-pass filters have frequency responses that permit a band of middle frequencies through, blocking low and high frequencies. Once you have identified the frequency response you desire for your filter, you must use filter design techniques to approximate that frequency response. There are families of IIR filters, such as Chebyshev and Elliptical, and they approximate the desired frequency response in different ways. Figure 29.4 presents two real IIR filter frequency responses. You can see how the Chebyshev and Elliptical filters approximate the ideal band-pass filter in different ways – the Chebyshev has more 'ripple' in the pass band with a flatter high and low response. The Elliptical filter has a flat pass band response, with more ripple in the low and high frequencies. You would choose an approach depending on your requirements.

That brief considering of filter design concludes our high-level introduction of IIR filters. In the following section, I will show you some simple filter design techniques based on the IIRCoefficients class in JUCE.

29.6 IIR filters in JUCE

Now that you have learned about IIR filters, I will show you how to implement some simple IIR filters in JUCE. In its IIRCoefficients class, JUCE includes some functions that allow you to compute IIR filter coefficients for common forms of frequency response such as low pass, high pass and so on. These are effectively a built-in 'filter design' system.

You can then use a similar method to the one you used in the JUCE convolution plugin to create a DSP chain containing a filter which uses those coefficients. The process is the same as for the JUCE convolution implementation from earlier. I will describe that process again here, with the adjustments needed to create the IIR plugin.

29.6.1 Prepare the plugin project

Make a copy of the plugin starter project described in the repo guide section 39.2.2. Load up the CMakeLists.txt file and edit the target name, which will start as 'minimal_plugin'. You might call it something with the words IIR and filter in it. Replace all instances of that target name in the CMakeLists.txt file. Also, set your PLUGIN_CODE to something unique.

Next, in the juce_add_plugin command, set 'IS_SYNTH' to 'FALSE'. This will allow the plugin to be an effects unit. Edit the target_link_libraries block by adding the 'juce::juce_dsp' module. My target_link_libraries block looks like this:

```
1  target_link_libraries(IIR_plugin
2      PRIVATE
3          juce::juce_audio_utils
4          juce::juce_dsp
5      PUBLIC
6          juce::juce_recommended_config_flags
7          juce::juce_recommended_lto_flags
8          juce::juce_recommended_warning_flags)
```

At this point, you should attempt to build the plugin. Depending on your IDE, take the appropriate steps to generate and build the IDE project.

29.6.2 Create the IIR DSP chain

In your project, open up the PluginProcessor.h file and add the following line to the private section of the Plugin class you find there:

```
1  juce::dsp::ProcessorChain<juce::dsp::IIR::Filter<float>>
       processorChain;
```

Now complete the housekeeping with the processorChain by implementing reset on your plugin. In PluginProcessor.h:

```
1  void reset() override;
```

Then, the implementation in PluginProcessor.cpp:

```
1  void TestPluginAudioProcessor::reset()
2  {
3      processorChain.reset();
4  }
```

29.6.3 Generate some coefficients

The final things to implement are prepareToPlay and processBlock. prepareTo-Play is the function called by the plugin host when it wants to tell the plugin what the audio configuration is. processBlock is the function the host calls whenever it wants the plugin to process a block of audio. In PluginProcessor.h, you should already have the declaration of these functions, so you can go straight to PluginProcessor.cpp and add their implementations. Firstly, prepareToPlay:

```
// put this in prepareToPlay:
const auto channels = jmax (getTotalNumInputChannels(),
getTotalNumOutputChannels());

//Tell the processor about the audio setup
processorChain.prepare ({ sampleRate,
(uint32) samplesPerBlock,
(uint32) channels });

// now set up the coefficients
auto& filter = processorChain.get<0>();
filter.coefficients = juce::dsp::IIR::
        Coefficients<float>::makeLowPass(sampleRate, 500, 0.1);

reset();
```

That code informs the processor chain about the sample rate, block size and channel count. Then, it generates some coefficients for a low pass filter using the statically callable makeLowPass function on the 'Coefficients' class. Statically callable means we can call the function directly without instantiating a Coefficients object first.

Now, put this into processBlock:

```
ScopedNoDenormals noDenormals;
//How many channels?
const auto numChannels = jmax (getTotalNumInputChannels(),
                               getTotalNumOutputChannels());

// convert the buffer we were sent by the plugin host
// (buffer) into an AudioBlock object that the
// convolution processor can understand:
auto inoutBlock = dsp::AudioBlock<float>
                (buffer).getSubsetChannelBlock (0, (size_t)
                numChannels);

// carry out the convolution 'in place.'
processorChain.process (
        dsp::ProcessContextReplacing<float> (inoutBlock));
```

Now compile, debug your errors and run it in your plugin host. The most obvious feature to add is a GUI to control the filter's features. You should be able to make calls from the GUI (PluginEditor) to the PluginProcessor to pass updated filter parameters. You could then store the updated parameters somewhere and then update the coefficients the next time processBlock is called. That way, you should not have a problem with the filter changing in the middle of a block being processed. See if you can implement it.

29.7   The connection with neural networks

Writing in a paper about neural effects, engineers from Native Instruments compare IIR filters to recurrent neural networks and FIR filters to convolutional neural networks[22].

29.8   Progress check

In this chapter, you have learned about infinite impulse responses. You should be able to explain why infinite impulse responses might be preferable to finite impulse responses. You should know what a frequency response is and be able to interpret frequency response plots. You should be able to implement an IIR filter that can process a WAV file in a command-line program. You should also be able to use JUCE's ProcessorChain functionality to implement IIR filters using preset coefficients in a JUCE plugin.

30

Waveshapers

In this chapter, I will guide you through the process of creating distortion effects using waveshaping. You will start by implementing simple clip, relu, and sigmoid waveshapers available in libtorch as command-line programs to demonstrate the basic concepts. Then, you will create a basic JUCE waveshaping plugin that allows you to reshape waveforms in real-time. At the end of the chapter, you will see how you can combine waveshaping with the other DSP techniques seen in previous chapters to construct a complete guitar amplifier emulator as a JUCE plugin.

30.1   What is a waveshaper?

FIGURE 30.1
Digital signal processing makes waveshaping much easier than it used to be.

In a 1979 Computer Music Journal article, computer music pioneer Curtis Roads who is the author of the infamously thick Computer Music Tutorial, which should proudly grace and bend all serious computer music programmers' bookshelves, wrote that "The terms 'non-linear distortion' and 'waveshaping' have yet to be explained in these pages". Roads describes waveshaping as the process of passing a signal through a typically non-linear transfer function. The transfer function dictates an output amplitude given an input amplitude. In figure 30.2 you can see how various waveshaping transfer functions look and how they affect a sine wave test signal. Note how the waveshaper can add additional frequency components to the signal. This is the essence of distortion and saturation effects. In programming terms, implementing a waveshaper should be as simple as iterating over the signal and calling a function to process each sample in the signal. Unlike IIR and FIR filters, it is not necessary to maintain a memory of previous inputs or outputs. The interesting part is what you put in the body of that transfer function.

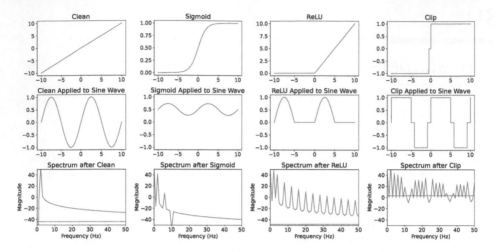

FIGURE 30.2

The effect of different waveshaper transfer functions on a sinusoidal signal. Top row: transfer functions, middle row: sine wave signal after waveshaping, bottom row: spectrum of waveshaped sine wave.

30.2 Simple C++ waveshaper implementation

Now we have learnt a little about the basic concept of waveshaping and seen how some transfer functions affect a signal we shall implement a simple C++ program that carries out waveshaping on a WAV file, saving the results to disk. First of all, here are the activation functions seen in figure 30.2:

```cpp
// includes to get some math and other operations
#include <algorithm>
#include <cmath>
// relu sets the value to zero
// if it is negative, otherwise,
// it leaves the value unchanged
float relu(float x) {
    return std::max(0.0f, x);
}

// sigmoid is a bit more subtle, applying
// a kind of DC offset combined
// with a compression effect at the extremes
// of the value range
float sigmoid(float x) {
    return 1.0f / (1.0f + std::exp(-x));
```

```
17 }
18
19 // clip pushes any values above a
20 // a threshold to a maximum value
21 float clip(float input, float clip_value) {
22     if (std::abs(input) > clip_value) {
23         return (input < 0) ? -1.0f : 1.0f;
24     }
25     return input;
26 }
```

Put together a program that uses the loadWav and saveWav functions we saw earlier to load a WAV, pass it through one of these waveshaping transfer functions and save it back out to disk as a new WAV file. Add the name of the function you applied to the name of the file. Here is a main function to get you started.

```
1 int main(){
2     std::vector<float> audio =
3         myk_tiny::loadWav("../../audio/drums_16bit.wav");
4     for (auto i=0;i<audio.size();++i){
5         audio[i] = relu(audio[i]);
6     }
7     myk_tiny::saveWav(audio, 1, 44100,
8                       "../../audio/waveshaped.wav");
9 }
```

You can find a working version of this waveshaper in the code repository section 39.5.12. Experiment with the different waveshapers and some different types of audio signals. How would you characterise the sound of the different transfer functions? For a challenge, can you think of a suitable transfer function to implement a rudimentary dynamic range compressor? Can you work out how to make the transfer function stateful so your compressor has an attack then a release time?

30.3   Basic JUCE waveshaper

Implementing a waveshaper in JUCE is fairly straightforward. In this section we will start with a 'DIY' JUCE waveshaper plugin because that will allow you to see exactly what is going on. In the next section, we will use the JUCE DSP processor chain components to rebuild the waveshaper as that allows us to integrate with the IIR and FIR implementations seen earlier.

Start with the plugin starter project described in section 39.2.2. Make the usual changes to the CMakeLists.txt file for the plugin and project properties. Make sure the juce_add_plugin block has IS_SYNTH set to FALSE. Now you just need to add the activation functions to the private section of your PluginProcessor.h:

```
1 float relu(float x);
2 float sigmoid(float x);
3 float clip(float x, float clip);
```

Then add the implementations to PluginProcessor.cpp. These are the same as the implementations in the command line program above but they are namespaced into the PluginProcessor class:

```
1 float TestPluginAudioProcessor::relu(float x){
2     ...
3 }
4 float TestPluginAudioProcessor::sigmoid(float x){
5     ...
6 }
7 float TestPluginAudioProcessor::clip(float x, float clip){
8     ...
9 }
```

Go ahead and replace the ... with the implementations of the transfer functions.

The final step is to use the transfer functions to process the audio in the PluginProcessor's processBlock function. Here is some minimal code to apply the clip waveshaper to the incoming samples. Put this into processBlock in PluginProcessor.cpp:

```
1 juce::ScopedNoDenormals noDenormals;
2 auto inChannels  = getTotalNumInputChannels();
3 auto outChannels = getTotalNumOutputChannels();
4
5 for (int c = 0; c < inChannels; ++c)
6 {
7     auto* cData = buffer.getWritePointer (c);
8     for (auto s=0;s < buffer.getNumSamples();++s){
9         cData[s] = clip(cData[s], 0.1);
10    }
11 }
```

You can find a working version of this plugin in the code repository section 39.5.13. Experiment by adding parameters controls to the user interface, for example, to vary the clip parameter. You can also add user interface controls to switch between different waveshaping transfer functions.

30.4   JUCE Waveshaper using ProcessorChain

Now you have a simple JUCE waveshaper plugin, we will do it all again but this time using the JUCE DSP processor chain. This method is a little more complex than the basic waveshaper from the previous section as it abstracts some of the process but this method has the advantage of easier integration with other JUCE DSP processor chain components such as IIR and FIR filters.

Here is an overview of the steps we will work through:

1. Prepare the plugin project, adding the JUCE DSP module to linker instructions

2. Add a juce::dsp::ProcessorChain to the plugin

3. Set up the ProcessorChain with a waveshaping function in the plugin constructor

4. Implement reset and prepareToPlay functions

5. Finally, pass the audio buffer to the ProcessorChain in the processBlock function of the plugin

30.4.1   Prepare the plugin project

Start a new plugin project from the basic plugin project described in code repository 39.2.2. Rename the project and set the plugin code in the CMakeLists.txt file, as usual. Add a line to the private section of the target_link_libraries command in CMakeLists.txt as follows, to include the JUCE DSP module in your project build:

```
target_link_libraries(myk_waveshaper
PRIVATE
    juce::juce_audio_utils
    juce::juce_dsp # this line!!!!
PUBLIC
    juce::juce_recommended_config_flags
    juce::juce_recommended_lto_flags
    juce::juce_recommended_warning_flags)
```

Verify that the default project builds and debug any issues as necessary.

30.4.2   Add and prepare a juce::dsp::ProcessorChain

Open the plugin processor header file PluginProcessor.h and add the following the private section of the class definition:

```
juce::dsp::ProcessorChain<juce::dsp::WaveShaper<float>>
    processorChain;
```

This adds a ProcessorChain to the plugin which it can use to carry out the waveshaping. Note that the code specifies the float data type for the chain. Now you need to configure the WaveShaper processor so it calls a particular transfer function. JUCE WaveShapers use lambda functions to carry out their processing. Lambda functions are similar to anonymous functions in Javascript – essentially they allow you to provide a block of code to execute when needed. The following code will set up the waveshaper with a transfer function that simply returns the input:

```
1  // setup the wave shaper
2  auto& waveshaper = processorChain.template get<0>();
3  waveshaper.functionToUse = [](float x){
4      return x;
5  };
```

If you have not seen a lambda (in C++) before, here is a quick breakdown of the parts:

```
1  [] // the [] allows you to pass in a
2      // 'scope' for the lambda, e.g. you
3      // can pass in 'this'
4      // and the lambda can
5      // see the PluginProcessor instance
6  (float x) // the next part specifies
7             // the parameters, in this case a float
8  {
9    // then the body of the
10   // function - the code that will execute
11     return x;
12 };// don't forget the semi colon
```

Over to you – can you take the code from the body of the clip function seen above and use it to replace the 'return x' statement from the lambda? You might wonder how to get access to a control parameter for the clip code, given the lambda only has a single argument. For now, hard-code it, and we will come back to that once we have the whole setup working.

30.4.3   Implement reset and prepareToPlay functions

Next, as you did in the IIR and FIR JUCE implementations seen earlier, you need to override the reset function (which is not done by default on the plugin project) such that it calls reset on the ProcessorChain. You also need to implement prepareToPlay such that it calls prepare on the ProcessorChain.

Add this to the public section of PluginProcessor.h to initiate an override of the reset function:

```
1  void reset() override;
```

Implement reset in PluginProcessor.cpp like this:

```
1  void TestPluginAudioProcessor::reset(){
2      processorChain.reset();
3  }
```

Implement prepareToPlay in PluginProcessor.cpp like this:

```
1  void TestPluginAudioProcessor::prepareToPlay
2      (double sampleRate, int samplesPerBlock)
3  {
4      // work out how many channels
5      const auto channels = jmax (getTotalNumInputChannels(),
6                              getTotalNumOutputChannels());
7      // send a special 'struct' to the
8      // processorChain with the necessary config:
9      processorChain.prepare ({ sampleRate,
10                              (uint32) samplesPerBlock,
11                              (uint32) channels });
12
13 }
```

Note how there is a slight oddness where you have to organise the parameters received by the prepareToPlay function and the channel information into a struct with three properties since processorChain.prepare expects it.

30.4.4   Implement waveshaping in processBlock

The next step is to pass audio through the ProcessorChain in the plugin's processBlock function. This is identical to the implementation in the IIR and FIR examples:

```
1  ScopedNoDenormals noDenormals;
2  const auto inChans  = getTotalNumInputChannels();
3  const auto outChans = getTotalNumOutputChannels();
4  const auto numChans = jmax (inChans, outChans);
5  auto inoutBlock = dsp::AudioBlock<float>(buffer)
6                      .getSubsetChannelBlock
7                      (0, (size_t) numChannels);
8  processorChain.process (
9              dsp::ProcessContextReplacing<float> (inoutBlock)
10             );
```

30.5   Parameterising the waveshaper

I referred to a problem earlier – the lambda is not set up to receive a control parameter. It just receives the sample from the buffer and processes it. How can we provide a parameter to the lambda to allow for dynamic control of the clip point? First you need to add the parameter to the plugin, so let's just do it simply as a public field. Having parameters as public fields is not exactly the neatest way to do things but it will enable a simpler implementation without getters and setters.

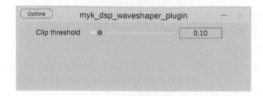

FIGURE 30.3
Automatically generated generic UI for the waveshaper plugin.

We will add the parameter as a real plugin parameter – in PluginProcessor.h's public section add the following:

```
juce::AudioParameterFloat* clipThreshold;
```

Now in the constructor, instantiate the parameter and register it to the plugin with 'addParameter':

```
addParameter (clipThreshold = new juce::AudioParameterFloat (
    "clipThresh",
    "Clip threshold",
     0.0f,
     1.0f,
     0.1f
));
```

Now make the parameter editable via an automatically generated UI by replacing the line in the createEditor defined in PluginProcessor.cpp with the following:

```
juce::AudioProcessorEditor* TestPluginAudioProcessor::createEditor()
{
    return new GenericAudioProcessorEditor(*this);
    //return new TestPluginAudioProcessorEditor (*this);
}
```

The GenericAudioProcessorEditor will automatically generate a UI based in the parameters you have registered for the plugin. Build and run it to test. Your user interface should look something like figure 30.3, with a single slider to control the clip threshold.

30.5.1 Fixing the lambda function

The next step is to fix the lambda function which presently has no way to access the clip threshold parameter. Change the specification of the processorChain variable in PluginProcessor.h to allow it to accept a 'std::function' as its lambda:

```
juce::dsp::ProcessorChain<juce::dsp::WaveShaper
    <float, std::function<float(float)>>> processorChain;
```

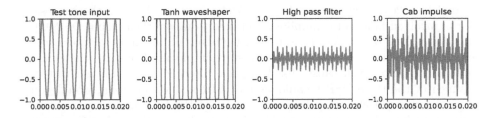

FIGURE 30.4

A sine wave passing through a series of blocks that emulate in a simplified way the processing done by a guitar amplifier.

Note how we added a second property to the WaveShaper template specification: std::function¡float(float). Previously we only specified one property – the type of the input to the waveshaping transfer function, float. Previously the waveshaper was using the default type of function for that second property which does not permit us to send in a 'scope'. Now we can update the code in the constructor in PluginProcessor.cpp with a different lambda:

```
auto& waveshaper = processorChain.template get<0>();
waveshaper.functionToUse = [this](float x){
    float clip = this->clipThreshold->get();
    if (std::abs(x) > clip) {
    return (x < 0) ? -1.0f : 1.0f;
    }
    return x;
};
```

Note how there is now a 'this' in the square brackets which provides access to the current plugin processor object to the lambda. This is the scope I mentioned earlier. Now the lambda can access the clipThreshold variable that we added to the public section of the plugin processor. This means the behaviour of the clip waveshaper can dynamically change. Get the plugin up and running then enjoy playing your favourite guitar, bass, synthesizer or even drums through it. I particularly enjoyed making a super-fuzzy bass guitar tone with it. You can find a working version of this plugin in the code repository section 39.5.14.

30.6 Working towards a guitar amp emulator

Now you have seen all the components you need to create a classical DSP-based guitar amp emulator, but how should you put them together to create a convincing emulation? In a 2013 white paper, Fractal Audio Systems describe the processing blocks involved in a guitar amplifier and how to model them using DSP[42]. Fractal, the makers of the Axe-Fx series of high-end guitar amp emulators, describe guitar amps as a series of filter and distortion blocks. Typically, the guitar goes into a pre-amp which causes fixed EQ-like filtering and non-linear distortion. Which of the three DSP techniques IIR, FIR and waveshaping would you use to model these blocks? Hopefully you answered IIR then waveshaper. Next, the signal goes into the tone section which is typically an adjustable EQ – more IIR filtering. Following that, the signal is distorted again as it is power-amplified – waveshaping. Finally the signal is converted to sound waves via a speaker and speaker cabinet. This last stage can be modelled by capturing an impulse response from the speaker and cabinet and using FIR and convolution techniques.

Implementing a full model of a guitar amp like the one described by Fractal Audio Systems is beyond the scope of this part of the book. This part of the book is intended to be a short introduction to DSP techniques in support of the later chapters about neural DSP. But frankly building amp models is a lot of fun, so here follows some instructions on how to build a simple version of a multt-stage amp model. It shows you how to construct a chain of JUCE DSP processors to carry out waveshaping, IIR filtering then convolution to model the pre-amp, tone stage and speaker cabinet, respectively.

30.6.1   Chaining JUCE DSP processors together

In the following instruction sequence you are going to see how to chain JUCE DSP processors together. You can find the complete project in the code repository section 39.5.15. The instructions cover the following steps:

1.   Create plugin project
2.   Define the processor chain
3.   Instantiate and prepare the processor chain
4.   Process audio with the processor chain

I suggest that a reasonable starting point is to make a copy of the waveshaper example you developed in the previous section. If you have heavily customised it, you might want to start from a cleaner example which you can find in the code repository section 39.5.14. Change the plugin name and ID. Build it to verify it works.

30.6.2   Prepare and instantiate the processor chain

Open PluginProcessor.h. If you started with my clean waveshaper example, you should see the follow declaration in the private section:

```
1  juce::dsp::ProcessorChain
2          <
3          juce::dsp::WaveShaper
4            <float, std::function<float(float)>>
5          > processorChain;
```

Note that I have broken it apart in the layout of the code so you can see that the processor chain presently consists of a single WaveShaper. Let's add some more processors. I am going to create the sequence of processors shown in figure 30.4. These blocks consist of a waveshaper to emulate an overdriven pre-amp, a high pass filter to emulate a simple tone effect and a convolution processor to emulate the behaviour of the speaker cabinet.

```
1  // set up types for filter module
2  using Filter = juce::dsp::IIR::Filter<float>;
3  using FilterCoefs = juce::dsp::IIR::Coefficients<float>;
4
5  juce::dsp::ProcessorChain
6      <
7      // stage 1: pre-amp distortion waveshaper
8      juce::dsp::WaveShaper
9      <float, std::function<float(float)>>,
10
11     // stage 2: tone control IIR filter
12     // wrapped in a duplicator to make
13     // it stereo
14     juce::dsp::ProcessorDuplicator<Filter, FilterCoefs>,
15
16     // stage 3: speaker cab emulattion
17     // FIR convolution
18     juce::dsp::Convolution
19     >
20     processorChain;
```

The code specifying the IIR filter is longer than the other two modules because the IIR filter cannot operate in stereo. So the IIR filter is wrapped in a ProcessorDuplicator which converts it to stereo.

We will need to be able to access the three items in the chain later and this is done via the get function and a numerical index. We saw this in the earlier code where we called get(0) on the processorChain to gain access to the waveshaper, but in that code the waveshaper was the only module. A convenient syntax for accessing the modules in the chain is to add this enum to the private section of PluginProcessor.h:

```
1  enum{
2      ws_index,
3      iir_index,
4      conv_index
5  };
```

That code creates a thing called ws_index with a value of 0, iir_index with a value of 1 and conv_index with a value of 2. You will be using those names in the next step. Now let's configure the processing blocks. In PluginProcessor.cpp, constructor section add this code, which sets up the waveshaper and the FIR convolver. Note that it does not set up the IIR filter yet because we need to know the sample rate to configure the coefficients on the IIR filter and we do not know the sample rate until prepareToPlay is called:

```
// setup the wave shaper
auto& waveshaper = processorChain.template get<ws_index>();
waveshaper.functionToUse = [this](float x){
    float clip = this->clipThreshold->get();
    if (std::abs(x) > clip) {
    return (x < 0) ? -1.0f : 1.0f;
    }
    return x;
};

// load the impulse response into the convolver.
auto& convolution = processorChain.template get<conv_index>();
juce::File impFile{"path/to/your/wav/here"};
convolution.loadImpulseResponse (
        impFile,
        juce::dsp::Convolution::Stereo::yes,
        juce::dsp::Convolution::Trim::no,
        0);
```

For brevity I removed some file checking code from the example above, but you can see it in full in the code repository section 39.5.15. To set up the IIR filter, add the following code to the prepareToPlay function in PluginProcessor.cpp:

```
// setup the IIR filter - in this case, a 200Hz high pass
auto& filter = processorChain.template get<iir_index>();
filter.state = FilterCoefs::makeHighPass (
                    getSampleRate(), 200.0f);

//Tell the processor about the audio setup
const auto channels = jmax (getTotalNumInputChannels(),
                    getTotalNumOutputChannels());
processorChain.prepare ({ sampleRate,
                    (uint32) samplesPerBlock,
                    (uint32) channels });
```

The first part configures the coefficients for a high pass 'tone' filter. The second part informs the processor chain how the audio is configured. Build and test.

30.6.3 Process audio with the processor chain

The final step is to pass incoming audio blocks to the processor chain in the processBlock function. This is exactly the same as in the other DSP module-based effects you have seen:

```
1  ScopedNoDenormals noDenormals;
2  const auto numChans = jmax (getTotalNumInputChannels(),
3                              getTotalNumOutputChannels());
4  auto inoutBlock = dsp::AudioBlock<float>(buffer)
5                    .getSubsetChannelBlock
6                       (0, (size_t) numChannels);
7  processorChain.process (
8           dsp::ProcessContextReplacing<float> (inoutBlock)
9           );
```

That code figures out the number of channels then uses the result to configure the audio data block for the processor chain, eventually calling process on the chain.

FIGURE 30.5
Capture an impulse response for the convolutional cabinet simulator.

Now build the project and enjoy playing your favourite instruments through it. There are all kinds of extensions you can carry out here. The first thing I did was to capture a real impulse response from a little 'Orange' guitar amp I had in my studio. To do this, I connected the output of my sound card to the aux input of the amp. This is the closest you can get to passing the signal cleanly through the speaker on this amp. I played a repeating click into the amp and recorded the output (the click playing through the amp) back into my sound card using a mic. You can see the setup in figure 30.5. Next I edited the recorded impulse response down to a single click and the response in an audio editor. I faded out the signal at the end. I then loaded the impulse response into the convolution module in the plugin. I was quite pleased with the result when I played a guitar through the setup. The impulse response is in the audio folder in the code repository for you to try. You can then explore the other components; for example, you could convert the high pass IIR filter into a peak filter and add some plugin parameters to control it, just as we did for the clip control. Following that, you can add more modules, for example, a FIR-based reverb at the end of the chain.

30.7 Progress check

In this chapter you have explored waveshaping and how it can be used to create distortion effects. You should now have a JUCE plugin that carries out waveshaping. At the end we saw how to combine waveshaping with the other DSP techniques of IIR and FIR filters into a guitar amplifier emulator.

31

Introduction to neural guitar amplifier emulation

In this chapter, I will give you a high-level overview of the process of creating a guitar amplifier emulation using a neural network. Through the process of training, neural networks can learn to emulate the complex non-linear behaviours of tube amplifiers and other guitar gear. I will explain key concepts like how training data guides the network parameters and how back-propagation enables the network to learn. At the end of the chapter, you will have developed a greater understanding of training a neural network to model a guitar amplifier, ready for the practical implementation in the following chapter.

31.1 From grey, white and black boxes to neural networks

In their review of Neural Network techniques for guitar amplifier emulation, Vanhatalo et al. describe three general approaches to guitar amp emulation: whitebox, grey-box and black-box[43]. White-box emulation takes account of the low-level behaviour of the electrical circuits involved. It includes techniques such as Wave Digital Modelling, which models the flow of current through circuit components[10]. On the other hand, black-box modelling is not concerned with the fundamental nature of the system; instead, it is only concerned with its response to a given input. Which category fits the DSP techniques we have been looking at? Take a moment to consider that. I would classify the convolutional reverb algorithm as a black box. This algorithm does not concern itself with the 'how' of natural acoustic spaces, merely the 'what' of their impulse responses. How about waveshapers and IIR filters?

Grey-box models sit between black- and white-box, typically consisting of several black-box models chained together. The selection of modules in a grey-box model is informed by the high-level components of the system it is trying to emulate. The three-part amp model you saw in the previous chapters is a grey-box model. Whilst no low-level circuit information is used, each module aims to emulate a specific, high-level part of a real amplifier, e.g. the pre-amp or the speaker cabinet.

There are established techniques for white, grey and black-box modelling, but they have limitations which neural networks can overcome. Typical black-box models are not necessarily able to cope with the highly non-linear behaviour of certain circuits found in guitar amplifiers, especially when they include valves. Low-level white-box circuit modelling, whilst it can model complex, non-linear signal processing behaviours, requires knowledge and understanding of the circuits in a given device, making its implementation complex and highly skilled. The resulting models might also be computationally expensive and challenging to deploy in a music technology context. Grey-box models are more manageable than white-box models to design as they abstract some of the details of the low-level circuitry but still require complex signal processing chains. Grey-box models can also suffer from the limitations of their black-box sub-components.

Neural networks are powerful black-box models that can model complex non-linear behaviours of many different circuits. So, they capture some of the advantages of white and black-box models. A fundamental feature of neural networks is that they learn to emulate circuits via a process called training. Given an input, such as a clean guitar signal and an output, such as a distorted guitar signal recorded from an amplifier, a neural network can learn the transformation from input to output and generalise it to any input. This is conceptually similar to the idea of convolution seen before, but it goes further because it can model a much greater range of audio transformations than simple convolution can.

31.1.1 Trainable amp modellers already exist

For completeness, it is worth noting that user-trainable modelling systems such as the Kemper Profiling Amplifier do exist. The Kemper does not presently use neural networks as far as I can tell; I presume it uses classical signal processing. But these kinds of amp modellers do work very well – according to a 2020 research study, even expert listeners find it difficult to differentiate between recordings made with modelled amplifiers and real guitar amplifiers[11]. However, the dynamic, phenomenological experiences of the guitar players making the recordings were not considered. Unlike the existing, non-neural network systems available today, neural networks are generally considered the next generation of amplifier modelling technology, even with robust proof of their superiority to existing modelling systems.

31.2   What is a neural network?

This book is available as a whole or as separate parts, and you can study the parts in any order as long as you complete the set-up instructions in part 1. So you may be reading this part of the book without having worked with neural networks before. The first thing you should know about the neural networks here is that they are essentially fancy versions of the DSP processes we have already seen. Like our DSP-based guitar amp emulator, neural networks take an input, process it, and produce an output. In fact, depending on the neural network architecture, the processing can involve previous inputs, previous outputs and non-linear distortion steps. Just like FIR, IIR and Waveshapers!

You may have heard of some neural network architectures such as recurrent networks (RNN), their descendant, the Long Short Term Memory network (LSTM) and convolutional networks, or CNNs. As I mentioned above, there is a connection between these architectures and traditional DSP techniques. As a team from Native Instruments who were modelling filters with neural networks put it, "convolutional neural networks (CNNs) can be thought of as a type of non-linear FIR filter, and recurrent neural networks (RNNs) as a type of non-linear IIR filter."[22]. To complete the trio, the neural network equivalent of Waveshaping is the activation function.

31.3   What is training?

FIGURE 31.1
Capturing training data from a guitar amplifier.

The big difference between classical DSP techniques and neural networks is how we figure out the parameter settings for the processing. Neural networks allow for very complex combinations of their basic units. I mentioned LSTMs and CNNs above, which are made from units similar to delays with feedback (IIR filters) and convolution processes (FIR filters), respectively. A deep network comprises multiple layers of processing units stacked up. Each of these units has a set of associated parameters that dictate how it operates. These are similar to the coefficients in FIR or IIR filters. However, traditional filter design techniques are inappropriate for neural networks due to their complexity and non-linearity.

Instead, we use a training program to find out the parameter settings for the neural network. The process of training we

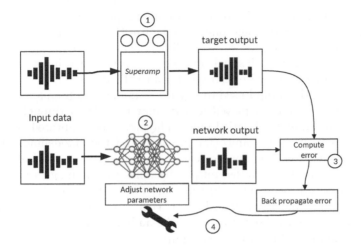

FIGURE 31.2
Four stages to train a neural network. 1: send the test input through the device (e.g. amp) you want to model, 2: send the test input through the neural network, 3: compute the error between the output of the network and amp, 4: update network parameters to reduce error using back-propagation. Back to stage 2.

will use can be categorised as supervised machine learning. Supervised because we tell the training program what the input and output should be. Machine learning because the training program and the neural network determine how to best achieve the desired output from the input.

So, figuring out the parameter settings for a neural network is called 'training'. Training is a bit like filter design, the process we discussed earlier for IIR filters wherein we work out the coefficients for the filter to achieve a desired frequency response. The difference is the way we go about finding the parameters. Figure 31.2 provides an overview of the four-stage neural network training procedure. The stages shown are:

1. Send the test input through the device you want to model and capture the output

2. Send the test input through the neural network

3. Compute the error between the output of the network and device you are modelling

4. Update the network parameters to reduce the error using back-propagation

5. Back to 2.

When training neural networks to process audio in specific ways, e.g. to emulate a guitar amp, we start with stage one from figure 31.2, preparing examples of inputs and the resulting outputs we would like. This is called the training data. Again, we are not too far from classical DSP here – capturing the input and output is similar to capturing an impulse response for an FIR filter.

Figure 31.1 shows the setup I used to capture training data from a small guitar amplifier I wanted to model. I played a particular test signal through the amp consisting of about a minute of varied guitar playing. I then captured the output either from the line-out of the amp (for pre-amp and tone modelling) or using the mic (for pre-amp, tone and speaker cab modelling). The test signal becomes the training input, and the recorded output becomes the training target. I want the neural network to process audio in the same way as the amp does. But wait – I should be careful what I say here. In fact, this is a black-box model, so I do not care how the amp goes about processing the audio. I only care about how the audio changes due to passing through the amp.

Stage two of the four-stage training process from figure 31.2 passes the training data inputs to the network and stores the outputs. Stage three compares them to the training data outputs and computes the error between the two. The training program uses a *loss function* to calculate the error between the output from the network and the correct output. There are many ways to measure the loss; for example, in audio applications, you might compare the spectrum of the desired output to the achieved output and calculate how different they are.

Once you have computed the 'error' using the loss function, stage four involves a famous algorithm called back-propagation. 'Back-prop' adjusts the parameter settings of the neural network – the loss is back-propagated through the network. Then, the training program returns to stage 2, passing the test signal into the network again. In this way, the network learns to approximate the transformation from the input to the output signal of the original amplifier.

Usually, the dataset is broken into *batches* and the network parameters are updated after each batch is passed through the network. So, as the training program works through the dataset, feeding it through the network and measuring the loss, it makes multiple adjustments to the neural network parameters. Once it has worked its way through the entire dataset, e.g. the 1 minute of audio in my guitar amp modelling example, we say that it has completed a training *epoch*. Training typically runs for several epochs.

As a side note, the idea of batch processing in parallel was one of the big ideas that kick-started the deep learning revolution in the noughties. Parallel batch processing involves running multiple instances of the neural network in parallel, passing a different input through each instance, and simultaneously computing the errors. Graphics Processors are very good at computing these kinds of parallel jobs. This parallel processing is what enables training over large datasets on large computing clusters. Luckily, amplifier modelling does not need a large dataset –

just one or two minutes of audio. So we can train our networks on more easily accessible hardware.

31.4   When is it trained?

To complete this preliminary discussion of training, we shall consider the question: 'When is it trained?'. For example, how many epochs do we need to run before the network is trained? As training proceeds, the loss across the dataset is reduced meaning the network gets better at emulating the guitar amplifier. If all goes well, at some point, the loss is low enough that you cannot hear the difference between the amplifier's output and the neural network's output. Great, but there is another aspect to this. What happens when the neural network processes a sound it has *never heard before*? Well, what would you like to happen? Imagine buying a reverb plugin to find out that it only works when you play a certain test signal through it, and not any other signal you might want. Similarly, if you are modelling a guitar amplifier, you need to know that your model can correctly process any signal, not just the test signal. In neural networks, this is called 'generalising'.

You test for generalising by keeping some of the dataset aside, say 10% or 20% of the data. You do not use this data for training; it is called the 'validation' data. You run the validation data through the network occasionally and measure the loss. This tells you how well the network performs on data it has not seen. You never back-propagate the validation error through the network. When the validation loss is low enough, training is complete. Sometimes, you find that the network is achieving a very low loss on the training data, but the validation data loss is going up. This means the network has zoned in too much on the training data and can no longer generalise. This is called 'overfitting' – the network is overfitting the training data to the point where it cannot fit the validation data. So you should stop training before the validation error starts ascending or when the error is not changing much between epochs.

31.5   Inference and deployment

Once you have trained your neural network to transform an audio signal like your target system, you can actually use the network to process signals. Running signals through the trained network without computing errors or adjusting parameters is called inference. Since one of the objectives of this book is to show you how to deploy trained neural networks in native applications and plugins, we need to think a bit about which tools and systems we will use to carry out training and inference.

Machine learning practitioners typically carry out network design, dataset preparation and training using a language like Python or Matlab. They may also deploy the models in Python, where deploying means making the network available for inference through some sort of service. However, we want to create audio software that regular music professionals can use without installing complicated machine-learning environments and without connecting to web-based services. We also want our networks to run on live audio signals in real time. So, we will use Python for training and C++ for inference. In the next chapter, you will find out how to take models created in Python and import them into your C++ programs.

31.6 Progress check

At this point, you should be able to differentiate between black-, grey- and white-box models and classify neural networks as black boxes. You should be able to make connections between common neural network architectures such as LSTM and ConvNets and classical DSP techniques such as IIR and FIR filters. You should have a general grasp of what is involved in training a neural network to model a guitar amplifier, i.e. creating training data and then iteratively updating the parameters of the network until it processes the audio in the correct way. You should know about the difference between validation data and training data. You should understand that training and inference are often done in different environments; we will train in Python and infer in C++.

32

Neural FX: LSTM network

In this chapter, you will put the previous chapter's theory into practice and start work on an LSTM model that can process audio signals. You will begin by setting up a Python environment suitable for creating and testing LSTMs. You will make some LSTMs and learn to pass audio signals through them in Python. Then, you will learn how to use TorchScript to export models from Python in a format that can be imported into a C++ program via libtorch. Once your LSTM is working in C++, you will see how to use it to process an audio signal read from a WAV file. At the end of the chapter, you will experiment with the performance of different-sized models to see which can be used in real-time scenarios.

32.1 The plan for this chapter

Before we start on the work in this chapter, you will need to set up your Python and C++ environments, including libtorch. Please refer to the instructions in Part I. From this point, I assume you have created a fresh Python virtual environment and are ready to build and run C++ programs.

In this chapter, we are going to carry out the following tasks:

1. Install some Python packages that allow us to work with neural networks and audio

2. Create a simple LSTM network in Python and examine how it processes audio

3. Export the LSTM network using TorchScript

4. Import the LSTM network into a minimal C++ program on the command line that passes a WAV file through the network

5. Convert the command-line C++ program into a VST plugin using JUCE

32.2 Installing Python packages for neural FX

You should have Python installed already. Create a new virtual environment for the work in this chapter:

```
1  # create the environment. Set nn-audio-venv
2  # to the complete path of the folder you want
3  # to create
4  # macOS/Linux
5  python3) -m venv nn-audio-venv
6  # Windows
7  python3.exe) -m venv nn-audio-venv
8  # activate the venv - macOS/Linux
9  source nn-audio-venv/bin/activate
10 # activate the venv - Windows
11 cd nn-audio-venv
12 .\Scripts\activate
```

If you use Anaconda, run the appropriate commands or use the UI to configure and activate your Anaconda virtual environment. Now, inside the virtual environment, install some packages – here are my preferred starter packages:

```
1  pip install torch librosa scipy numpy ipython jupyter
```

32.3 Pass numbers through an LSTM

Next, we will define and pass audio through a simple LSTM network. Here is a Python script that imports the torch module, creates an LSTM layer, and then passes a number through it:

```
1  import torch
2
3  # set a random seed so it generates the
4  # same network parameters
5  torch.manual_seed(10)
6  # make a simple LSTM network layer
7  my_lstm = torch.nn.LSTM(1, 1, 1)
8  # make an input - note is a 2D structure [[]]
9  in_t = torch.tensor([[1.0]])
10 # pass it through the LSTM layer
11 out_t,hx = my_lstm.forward(in_t)
12 # print the results
13 print(out_t)
```

Go ahead and run the script. Here is what I see when I run it:

```
1 tensor([[0.0318]], grad_fn=<SqueezeBackward1>)
```

That number is the output from the LSTM in response to the 1.0 we sent in. What about the other variable that came back, hx? That is the hidden state of the LSTM after we passed in the 1.0. Think of the LSTM like an audio delay effect – after you pass in a signal, the delay effect holds on to that signal and repeats it. So, the delay effect changes its state after you pass a value in, and so does the LSTM. The PyTorch LSTM layer is designed so you can choose to reset it or keep its previous state each time you pass in a number. So, if you hold on to that hidden state value hx, you can pass it into the network next time, and it will remember what state it was in after receiving the first value.

We will come back to that feature later. Right now, I want you to see what happens when you pass a longer signal into the network. You can actually pass a sequence of values in and receive the output sequence all at once. Try this in a fresh script:

```
1 import torch
2 torch.manual_seed(10)
3 my_lstm = torch.nn.LSTM(1, 1, 1)
4
5 # make an input
6 in_t = torch.tensor([[1.0], [0], [0], [0], [0], [0]])
7 # pass it through the LSTM layer
8 out_t,hx = my_lstm.forward(in_t)
9 # print the results
10 print(out_t)
```

Do you remember what the name is for that particular signal? Hopefully, you said 'the impulse signal'. I see the following result:

```
1 [[0.0318],
2 [0.1224],
3 [0.1459],
4 [0.1563],
5 [0.1615],
6 [0.1642]]
```

Interesting! Does the signal ever die off? Try this:

```
1 import torch
2 torch.manual_seed(10)
3 my_lstm = torch.nn.LSTM(1, 1, 1)
4
5 # 100 zeros
6 in_t = torch.zeros((100,1))
7 # put 1.0 at the start
8 in_t[0][0] = 1.0
9
10 out_t,hx = my_lstm.forward(in_t)
11 print(out_t)
```

When I run that script, I see the output pegging at 0.1678 after about 20 zeroes have gone through the network. To me, this looks like a DC offset. The network responds to the signal and eventually settles down, but with a constant DC offset. Try commenting out the call to manual_seed so the network is random each time you run the script. Run the script a few times. The network always settles on a constant, but each time it is different.

32.4   What is a tensor?

At this point, you have seen the word tensor used several times. What is a tensor? I like to think of certain data structures in a sequence of increasing complexity. The sequence goes: scalar, vector, matrix, tensor:

$$0.5 \rightarrow scalar$$

$$\begin{bmatrix} 0.5 & 0.1 & 0.25 \end{bmatrix} \rightarrow vector$$

$$\begin{bmatrix} 0.5 & 0.1 & 0.25 \\ 0.05 & 0.9 & 0.4 \end{bmatrix} \rightarrow matrix$$

$$\begin{bmatrix} \begin{bmatrix} 0.5 & 0.1 & 0.25 \\ 0.05 & 0.9 & 0.4 \end{bmatrix} & \begin{bmatrix} 0.5 & 0.1 & 0.25 \\ 0.05 & 0.9 & 0.4 \end{bmatrix} \end{bmatrix} \rightarrow tensor \qquad (32.1)$$

This is taking the software engineering definition of a tensor which defines it as a regular-shaped structure of numbers. We can actually wrap any of those shapes in a PyTorch tensor though:

```
1  t1 = torch.tensor(1.0)
2  t2 = torch.tensor([0.5, 0.1])
3  t3 = torch.tensor([[0.5, 0.1, 0.25],
4                     [0.05, 0.9, 0.4]])
5  t4 = torch.tensor([
6                     [[0.5, 0.1, 0.25],
7                      [0.05, 0.9, 0.4]],
8                     [[0.5, 0.1, 0.25],
9                      [0.05, 0.9, 0.4]],
10                    ])
```

Tensors are the currency of neural networks. Everything needs to be converted into some sort of tensor to work in neural network-world.

32.5   Pass audio through an LSTM

At this point, I am keen to hear what it sounds like when you send a signal through
the network and what it sounds like if you make an LSTM layer with more units.
So, let's experiment with some networks. Here is some code that synthesises a sine
tone at 400Hz with a sample rate of 44,100Hz, passes it through an LSTM, and
then saves both the original sine and processed sine to disk as WAV files:

```
import torch
import numpy as np
from scipy.io.wavfile import write

torch.manual_seed(10)

# synthesize a 400Hz sine at 44,100Hz sample rate
freq = 400
clean_sine = np.sin(
     np.arange(
          0, np.pi * freq*2,
          (np.pi * freq * 2) /44100),
          dtype=np.float32)

# reshape it so each sample is in its own
# box [[], [] ... []]
clean_sine = np.reshape(
               clean_sine,
               (len(clean_sine), 1))

my_lstm = torch.nn.LSTM(1, 1, 1)
in_t = torch.tensor(clean_sine)

# pass it through the LSTM layer
out_t,hx = my_lstm.forward(in_t)
# print the results
print(out_t)

write('sine_400.wav', 44100, clean_sine * 0.5)
write('sine_400-lstm.wav', 44100, out_t.detach().numpy())
```

Most of the work in that script was synthesising the sine tone (lines 8–13) and
reshaping it for input to the LSTM (lines 17–19). You can see the complete script
in the code repository section 39.5.16.

Now, let's analyse what the LSTM did to the sine wave regarding the time and
frequency domain of the signal. Figure 32.1 shows the 400Hz sine tone before and
after it passed through the LSTM in the time and frequency domains. You can see
that the sinusoidal shape has been 'pinched' at its low point and widened at its
high point. This results in the creation of additional harmonics. So you can see that

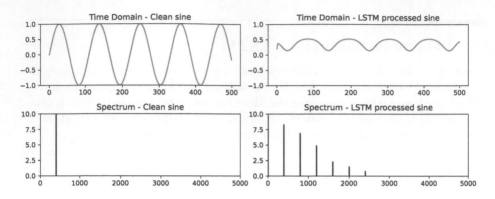

FIGURE 32.1
What does our simple, random LSTM do to a sine wave? It changes the shape of the wave and introduces extra frequencies.

even with one LSTM unit, we are already creating the kind of harmonic distortion that guitar players wax lyrical about when discussing their valve amplifier setups.

32.6   What is the LSTM doing?

As we saw in the previous section, a single LSTM layer with a single unit can add interesting harmonic distortion to audio signals. How does it do that? An LSTM unit is a rather elaborate delay and feedback unit with four distinct, parameterised components: input gate, forget gate, output gate and update system. These components work together to manage previous inputs, previous state and previous outputs to allow the unit to compute its subsequent output. As the name suggests, it can take account of the short-term, i.e. recent inputs and outputs, and the longer-term, i.e. less recent inputs and outputs. Hochreiter and Schmidhuber invented the LSTM in 1997 to solve the problem of networks losing track of older parts of the signal they were modelling[17]. The design allows the LSTM to learn to deliberately forget certain things and to remember others.

Let's take a quick look at the parameters that control those four components:

```
for p in my_lstm.parameters():
    print(p.data)
```

I see the following output:

```
tensor([[-0.5418],
```

```
 2      [-0.3582],
 3      [-0.0454],
 4      [ 0.2963]])
 5 tensor([[-0.9843],
 6      [-0.2911],
 7      [ 0.1443],
 8      [ 0.1551]])
 9 tensor([-0.9655,   0.1192,  -0.1923,   0.2199])
10 tensor([ 0.6780,   0.5214,  -0.7319,  -0.6145])
```

I counted four distinct parameter tensors there, but each parameter is made of four numbers. So, for an LSTM layer with a single unit, we have 16 trainable parameters. As noted above, the LSTM has four distinct components, and those four sets of numbers dictate the behaviour of each of those components, eventually resulting in the LSTM subtly remembering some things and forgetting others. As the number of units in the layer increases, the number of parameters increases, and not in a linear way, as each internal parameter interacts with several of the others:

```
1 Units: 1 params: 16
2 Units: 2 params: 40
3 Units: 3 params: 72
4 Units: 4 params: 112
```

At this stage, it is not necessary to go into more detail about how LSTMs work; having a concept of LSTMs as clever memory devices with adjustable parameters is sufficient. Now you should experiment with passing signals through different-sized LSTM networks. If you use the seed setting capability you can restore networks on subsequent runs if you find any that do interesting things to the signal. You will find a helpful notebook I provided – check out section 39.5.17 in the repo guide.

32.7   Exporting an LSTM model with TorchScript

You will now work towards running an LSTM in a C++ program. The first step is to export the LSTM from Python to C++. If you have seen the meta-controller example elsewhere in the book, you will know that you can design and train neural networks entirely in C++ using libtorch. In that setup, no Python was needed. It made sense to avoid Python as, for that example, we wanted the end-user to be able to interactively train the neural network within their DAW software. So, everything needed to run inside a compiled plugin.

Here, I will use a different workflow where you design and train the network in Python and then import the trained model into C++. Why would you do this instead of just working in C++? One reason is that Python is a more rapid language for prototyping, especially if you use notebooks to interactively develop your code. This means you can quickly hack around and get your data into the right shape, and so on. Another reason to use Python for design and training is that you can easily plot signals and other graphs in Python. But the main reason for our purposes is the ease with which you can access GPU accelerated training from Python, be it your own, local GPU or a remote GPU, e.g. running in Google's compute infrastructure. For the guitar amp emulator system, we do not intend for end-users to carry out training, so we should design the workflow to make our job as easy as possible.

So, if we are to have two environments, we need a way to get the neural network models from Python to C++. The method we will see uses TorchScript. Torch-Script is a format for saving a model that can be understood by both Python's PyTorch and C++'s libtorch environments. Let's go straight in and export our simple LSTM 'harmonic distortion' effect to TorchScript.

```
import torch
torch.manual_seed(10)
my_lstm = torch.nn.LSTM(1, 1, 1)
traced_lstm = torch.jit.trace(
                    my_lstm,
                    torch.rand(1, 1)
              )
traced_lstm.save('my_lstm.pt')
```

That's pretty straightforward. We have to pass some data through the network using the trace function, and then we receive a traced network, which we can save to a TorchScript (.pt) file.

32.8   Importing TorchScript models into C++

To load the exported TorchScript file into a C++ program, start with the minimal libtorch example from the code repository section 39.2.10. Edit the CMAKE_PREFIX_PATH property in CMakeLists.txt to point at the location of your libtorch folder. Try a test build to verify things are working ok. If you can build and run the program, you are ready to import from the TorchScript file. Edit main.cpp in the src folder:

```
//Load the model
torch::jit::script::Module my_lstm;
my_lstm = torch::jit::load("../my_lstm.pt");

//Create the impulse response signal
// and convert to an IValue vector
torch::Tensor in_t = torch::zeros({10, 1});
```

```
 8  in_t [0][0] = 1.0;
 9  std::vector<torch::jit::IValue> inputs;
10  inputs.push_back(in_t);
11
12  //Pass the data through the model
13  // store it as an IValue
14  torch::jit::IValue out_ival = my_lstm.forward(inputs);
15  std::cout << out_ival << std::endl;
```

This code replicates the Python example we saw earlier, which passes the impulse signal through the network. One oddity of imported TorchScript models which makes them different from directly defined models is the need to pass in the tensor as an IValue vector. IValues provide a generic way to pass data into Torch-Script models. The models also return IValue data but not a vector of IValues.

Here is the first part of the output on my machine:

```
1  ( 0.0318
2    0.1224
3    0.1459
4    ...
```

This is great as it looks very similar to the output received from the original model running in Python. The complete, working code is in the repo guide section 39.5.18.

32.9 Processing a WAV file with an LSTM in C++

Let's crack on and see if we can pass a WAV file through the LSTM in C++. Quite a few steps are involved in processing the data, as shown in figure 32.2. You will need to set up your project to include the tinywav library. You can see this section's complete, working code in the repo guide section 39.5.19. I assume the tinywav folder is located one folder above the folder where your project's CMakeLists.txt file is located. Update the add_executable command in the CMakeLists.txt file so it references your main.cpp file and the tinywav files:

```
1  add_executable(minimal-libtorch src/main.cpp
2      ../tinywav/tinywav.c
3      ../tinywav/myk_tiny.cpp)
```

Then edit the includes in your main.cpp to include the tinywav header:

```
1  #include "../../tinywav/myk_tiny.h"
```

Run the build process to verify your build is working. Then, add these lines to your main function to 1) load a WAV file, 2) convert it to a tensor, 3) reshape it, and 4) convert it to an IValue vector.

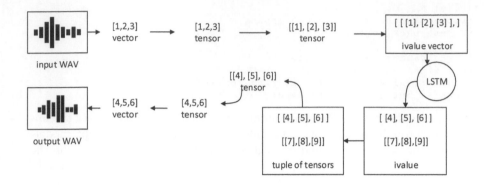

FIGURE 32.2
The steps taken to process a WAV file with a neural network through various shapes and data formats.

```
 1  // 1) get the WAV data as vector of floats
 2  std::vector<float> signal = myk_tiny::loadWav("mywav.wav");
 3
 4  // 2) convert float vector to tensor
 5  // without re-allocating memory
 6  torch::Tensor in_t = torch::from_blob(signal.data(),
 7                                        {static_cast<int64_t>
 8                                         (signal.size())});
 9
10  // 3) reshape tensor from [1,2,3] to [[1], [2], [3]]
11  in_t = in_t.view({-1, 1});
12
13  // 4) convert tensor to value vector
14  std::vector<torch::jit::IValue> inputs;
15  inputs.push_back(in_t);
```

Refer back to figure 32.2 for the steps leading up to entry to the LSTM to clarify what form the data is in each stage. Run that code; if it works, you have prepared the data for processing by the network. The most likely problem you will encounter is having the wrong filename for the WAV file or even the model – try using an absolute file path instead of a relative one.

Now, to feed the data to the neural network and save the result to a new WAV file:

```
 1  // feed the inputs to the network
 2  torch::jit::IValue out_ival = my_lstm.forward(inputs);
 3
 4  // convert the return to a tuple, then extract its elements
 5  auto out_elements = out_ival.toTuple()->elements();
 6
```

```
 7  //Take the first element of the tuple and convert it to a tensor
 8  torch::Tensor out_t = out_elements[0].toTensor();
 9
10  // reshape the tensor from [[1], [2], [3]] to [1,2,3]
11  out_t = out_t.View({-1});
12
13  // convert it to a vector of floats
14  float* data_ptr = out_t.data_ptr<float>();
15  std::vector<float> data_vector(data_ptr, data_ptr + out_t.numel());
16
17  // save it out to a file
18  myk_tiny::saveWav(data_vector, 1, 44100, "test.wav");
```

Again, refer back to figure 32.2 to see how the data is transformed at each step. If you are an experienced C++ developer, your C++ 'spidey-senses' might be telling you to worry about the memory allocation that might be going on during this process. You will see from my comments in the code that the main place I make an effort with memory allocation is the conversion to and from 'libtorch world'. So, I avoid copying the vector when I convert it to a tensor at the start, and I avoid copying the tensor when I convert it to a vector for saving at the end. Memory efficiency is a challenging and deep subject when working with libtorch. For example, how do you know that those internal processes in the LSTM model are memory efficient when the model was initially written in Python and is opaque from the C++ side? This is quite a deep topic, and we will see how to use a more efficient alternative to libtorch called RTNeural later. My take on optimisation is that it generally makes the code more complex and difficult to maintain and understand. Therefore, I try not to do anything silly in my code that really slows things down, but I prefer simple, understandable code. I work on the basis that optimisation is only necessary when the software does not run fast enough for the task at hand. So, let's test how fast the code runs.

32.10   Performance test

I created a performance testing program to see how fast we can run audio through a neural network. It is in the repo guide section 39.5.20. There is a Python script that generates LSTM models with exponentially increasing numbers of hidden units:

```
1  import torch
2  torch.manual_seed(10)
3  for i in range(8):
4      # Increase network hidden size exponentially
5      # so 1,2,4...128
6      my_lstm = torch.nn.LSTM(1, pow(2, i), 1)
```

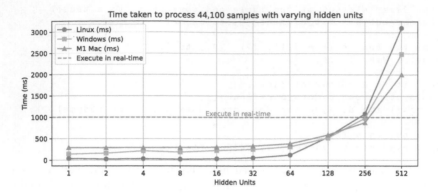

FIGURE 32.3

Time taken to process 44,100 samples. Anything below the 1000ms line can potentially run in real-time. Linux seems very fast with low hidden units, but Windows and macOS catch up at 128 units.

```
7    traced_lstm = torch.jit.trace(
8                    my_lstm,
9                    torch.rand(1, 1)
10                )
11   fname = 'my_lstm_' + str(pow(2, i)) + '.pt'
12   print("saving ", fname)
13   traced_lstm.save(fname)
```

Then, there is a C++ program defined in main.cpp in the example that runs a one-second test signal through each of the exported models and times it using the std chrono library. I will not go into the details of the code here as it is very similar to the previous code example. You should open up main.cpp and take a look. You will see that it times the complete process, starting from having a signal in a float vector through to receiving a vector of floats representing the signal after passing through the LSTM via the processes shown in figure 32.2. Figure 32.3 shows the results of running the program on my development machines: an Intel 10th gen Linux machine, an Intel 10th gen Windows machine and an M1 Mac Mini. The Linux machine can run up to 256 LSTM units in real-time. The Mac Mini can run up to 128 units in real-time, but it seems to have a heavier overhead on all sizes than the Intel CPU machine – even one LSTM unit takes 658ms on the Mac. The Mac is running a self-compiled libtorch 2.1. In a real-world scenario, there would be more processing going on here, but this simple program gives us an insight into the kind of networks we can expect to use in a real-time audio application. We will revisit the question of performance later in this part of the book when we deploy trained LSTM in plugins using libtorch and RTNeural.

32.10.1 Gradient calculation

The last concept I would like you to know about is the idea of computing gradients. This will be much clearer once you have learned about training neural networks in a later chapter, but it is relevant to performance, so I am mentioning it now. When you pass data through a network, the libtorch library does not just compute the output; it also computes gradients which describe how that data flows through the network. Those gradients are used during training to allow the back-propagation algorithm to update the parameters. We are not training right now – we only want to do inference. Inference, as I mentioned above, is when you pass data through the network to get the output. You can switch off gradient calculation when carrying out inference, and this uses fewer CPU cycles. Surprisingly, I found that this improved Windows and macOS performance but did not affect Linux performance. To turn off gradient calculation, put the following line ahead of your inference, which informs the libtorch system not to compute gradients:

```
torch::NoGradGuard nograd;
```

32.11 Progress check

Wow, that was a lot of number-crunching. At this point, you should be able to define a simple LSTM model in Python and then run signals through the model, plotting the results. You should then be able to export your model using TorchScript and import the model file to C++. With that imported model, you should be able to process a WAV file and save the result to a new WAV file. You should also be able to reflect on the performance of differently sized LSTM models and how that relates to their potential use in real-time scenarios.

33

JUCE LSTM plugin

In this chapter, you will build on your work with LSTMs in the previous chapter, eventually wrapping up the LSTM model in a JUCE plugin that can pass audio from the plugin host through the LSTM in real-time. First, you will convert your WAV processing C++ program to a block-based model. You will learn that block-based processing causes a problem wherein the state of the LSTM is reset, leading to audible glitches in the sound. You will learn how to solve the problem by retaining the state of the LSTM between blocks. This will involve returning to the Python code to trace the model with more parameters and learning which data structure to use to represent the LSTM state in your C++ program. Once the command-line WAV processing program works without generating artefacts, you will convert that program into a JUCE plugin and see how you can process audio received from a plugin host through an LSTM model.

33.1   Block-based signal processing

So far, we have passed audio signals through the LSTM in one large block. Under the hood (inside libtorch), the LSTM processes the audio sample by sample. We can pass it an audio block of any length, and it should cope with it. This is useful as plugins process the incoming signal in short blocks of 16 to 4096 samples, depending on the audio device settings. Let's start by changing the command-line C++ program to process the audio in blocks instead of in one shot. Go back to example 39.5.19 from the repo guide, which shows how to load a WAV file and pass it through an LSTM model. It goes something like this:

```
1  // load the TorchScript model
2  torch::jit::script::Module my_lstm = torch::jit::load('lstm.pt');
3  // load wav into float vector
4  std::vector<float> signal = myk_tiny::loadWav("mywav.wav");
5  // convert to tensor, change shape, then to value vector
6  torch::Tensor in_t = torch::from_blob(signal.data(), {static_cast<
       int64_t>(signal.size())});
7  in_t = in_t.View({-1, 1});
8  std::vector<torch::jit::IValue> inputs;
```

```
 9  inputs.push_back(in_t);
10  // pass it through the network:
11  torch::jit::IValue out_ival = my_lstm.forward(inputs);
```

If you want to convert that to block-based processing, how would you do it? If you are a seasoned programmer, you probably imagined some sort of iterator jumping by a block size, cutting out a block from the overall signal vector and passing it over for processing. That's my approach. Here is a function that processes a block of audio in this way:

```
 1  void processBlock(torch::jit::script::Module& model, // LSTM
 2          std::vector<float>& inBlock, // incoming signal
 3          std::vector<float>& outBlock) // outgoing signal
 4  {
 5      // convert to tensor without copy
 6      torch::Tensor in_t = torch::from_blob(
 7          inBlock.data(), {static_cast<int64_t>(inBlock.size())}
 8      );
 9
10      // reshape [1.2.3] to [[1], [2], [3]]
11      in_t = in_t.View({-1, 1});
12      // convert to value vector
13      std::vector<torch::jit::IValue> inputs;
14      inputs.push_back(in_t);
15
16      // pass-through model
17      torch::jit::IValue out_ival = model.forward(inputs);
18
19      // extract output
20      auto out_elements = out_ival.toTuple()->elements();
21      torch::Tensor out_t = out_elements[0].toTensor();
22      out_t = out_t.View({-1});
23
24      // convert to vector without copy
25      float* data_ptr = out_t.data_ptr<float>();
26      std::vector<float> data_vector(
27          data_ptr, data_ptr + out_t.numel());
28
29      // copy to outBlock - should not require memory allocation
30      if(data_vector.size() == outBlock.size()){
31          std::copy(data_vector.begin(),
32                  data_vector.end(),
33                  outBlock.begin());
34      }
35  }
```

Can you implement a loop in the main function that sends the complete audio file block by block to the processBlock function? Here is my solution – yours might be different.

```
 1  //Load the model
 2  torch::jit::script::Module my_lstm
 3          = torch::jit::load("../my_lstm.pt");
```

```
 4  // load WAV
 5  std::vector<float> signal
 6          = myk_tiny::loadWav("../../audio/sine_400_16bit.wav");
 7  // setup vector to store processed signal
 8  std::vector<float> outSignal(signal.size());
 9
10  // setup blocks for processing
11  int blockSize = 1024;
12  std::vector<float> inBlock(blockSize);
13  std::vector<float> outBlock(blockSize);
14  // loop through jumping a block at a time
15  for (auto s=0;s + blockSize < signal.size(); s += blockSize){
16      // copy signal into inBlock
17      std::copy(signal.begin() + s,
18                signal.begin() + s + blockSize,
19                inBlock.begin());
20      processBlock(my_lstm, inBlock, outBlock);
21      // copy outBlock to outSignal
22      // (won't need to do that in a real-time situation)
23      std::copy(outBlock.begin(),
24                outBlock.end(),
25                outSignal.begin() + s);
26  }
27  // save to WAV
28  myk_tiny::saveWav(outSignal, 1, 44100, "test.wav");
```

Can you test the program with some audio files? I am testing it with a 400Hz sine wave tone – I suggest you do the same and listen to the output carefully. You can find a fully working version of the block-based program in the repo guide section 39.5.21.

33.2   The problem of LSTM state

If you listen carefully to the output from the block-based program, you might hear small discontinuities in the signal. The discontinuities happen each time a new block is processed. I have created a plot showing a signal processed in a single block and multiple blocks in figure 32.2. You can clearly see the signal dropping off at regular intervals in the block-based plot. This is because each time you pass a block through the LSTM, the internal state of the LSTM is reset. Remember that the LSTM is a fancy delay line – it is as if the memory of the delay line has been wiped, and it takes a little time to fill up again. The numerical difference is most aptly illustrated with a simple Python script:

```
1  import torch
2
3  torch.manual_seed(21)
```

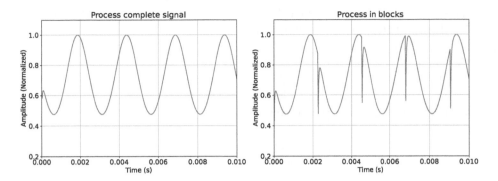

FIGURE 33.1
Block-based processing leads to unwanted artefacts in the audio. The left panel
shows the output of the network if the complete signal is processed in one block.
The right panel shows what happens if the signal is passed through the network
in several blocks. The solution is to retain the state of the LSTM between blocks.

```
4  my_lstm = torch.nn.LSTM(1, 1, 1)
5
6  # pass in 0.5
7  input = torch.tensor([[0.5]])
8  output, _ = my_lstm.forward(input)
9  print(output)
10 # pass in 0.1
11 input = torch.tensor([[0.1]])
12 output, _ = my_lstm.forward(input)
13 print(output)
14 # pass in 0.5 then 0.1 in one sequence
15 input = torch.tensor([[0.5], [0.1]])
16 output, _ = my_lstm.forward(input)
17 print(output)
```

The results I see are:

$$0.5 \rightarrow 0.0575$$
$$0.1 \rightarrow 0.0836$$
$$[0.5, 0.1] \rightarrow [0.0575, 0.1191] \tag{33.1}$$

The crucial thing to notice is that 0.1 produces a different output when it
comes after 0.5 in a single block. This tells you that the LSTM is stateful – when
you pass in a block, it runs through each number in turn but remembers its state
between them. At the end of the block, it resets its state. Luckily, it is possible
to store the state then to send it back next time you call 'forward'. Notice how I

store the second value returned from the call to forward and then pass it into the second call to forward in the following code:

```
1  input = torch.tensor([[0.5]])
2  output, state = my_lstm.forward(input)
3  print(output)
4
5  input = torch.tensor([[0.1]])
6  output, state = my_lstm.forward(input, state)
7  print(output)
8
9  input = torch.tensor([[0.5], [0.1]])
10 output, _ = my_lstm.forward(input)
11 print(output)
```

This time, the result from the individually passed values is the same as the values passed in a block:

$$0.5 \to 0.0575$$
$$0.1 \to 0.1191$$
$$[0.5, 0.1] \to [0.0575, 0.1191] \tag{33.2}$$

So, the secret to retaining the state between calls to forward is to store the returned state and send it back to the next call to forward. We shall see shortly how to create the state data in the appropriate form in C++, which is not a well-documented operation in libtorch for C++ but is well described in the LSTM examples for Python. Before we do that, though, we need to fix the TorchScript model.

33.3 Updating the TorchScript model to receive a state value

The original code we used to generate the TorchScript model looks like this:

```
1  torch.manual_seed(21)
2  my_lstm = torch.nn.LSTM(1, 1, 1)
3  traced_lstm = torch.jit.trace(my_lstm, (torch.rand(1, 1),))
4  traced_lstm.save('my_lstm.pt')
```

When we trace the model, we simply pass it an input, not any state data. Will the resulting TorchScript model accept state data? Let's try it:

```
1  torch.manual_seed(21)
2  my_lstm = torch.nn.LSTM(1, 1, 1)
```

```
 3 traced_lstm = torch.jit.trace(my_lstm, (torch.rand(1, 1),))
 4 traced_lstm.save('my_lstm.pt')
 5
 6 # generate some state data
 7 input = torch.tensor([[0.5]])
 8 output, state = my_lstm.forward(input)
 9 print(output)
10 # pass the state data to the traced TorchScript model
11 output, state = traced_lstm.forward(input, state)
12 print(output)
```

This code crashes on my machine with the following error: RuntimeError: forward() expected at most 2 argument(s) but received 3 argument(s). Declaration: forward(__torch__.torch.nn.modules.rnn.LSTM self, Tensor input) – ((Tensor, (Tensor, Tensor))).

Let's break that error apart. The error mentions the difference between the declared 'forward' function arguments and the arguments we tried to pass. It expects two, but we passed three arguments. The first argument is called 'self', which is always the first argument when using object-oriented Python and is automatically passed. The second argument is a Tensor, which is the input. We passed a third argument: a (Tensor,Tensor), a tuple of tensors. That is the state value. The traced model does not expect a third argument. To update the traced model to accept this third state argument, we need to pass that argument when we trace it:

```
 1 import torch
 2 # network parameters
 3 input_size = 1 # mono signal
 4 hidden_size = 1
 5 num_layers = 1
 6 torch.manual_seed(21)
 7 my_lstm = torch.nn.LSTM(input_size, hidden_size, num_layers)
 8 input = torch.tensor([[0.5], [0.1]])
 9 h0 = torch.rand(num_layers, hidden_size)
10 c0 = torch.rand(num_layers, hidden_size)
11
12 # trace with input and state arguments
13 traced_lstm = torch.jit.trace(my_lstm, (input, (h0, c0)))
14 traced_lstm.save('my_lstm.pt')
15
16 # Now this works:
17 output, state = traced_lstm.forward(input, (h0,c0))
18 print(output)
```

In this case, I generated random data for inputs and state. Then, I used those values in the trace. I now have a TorchScript model that is ready to receive state data. Debugging models like this in Python is much easier than in C++, as the errors are a little clearer, and the debugging cycle is faster. I know this because

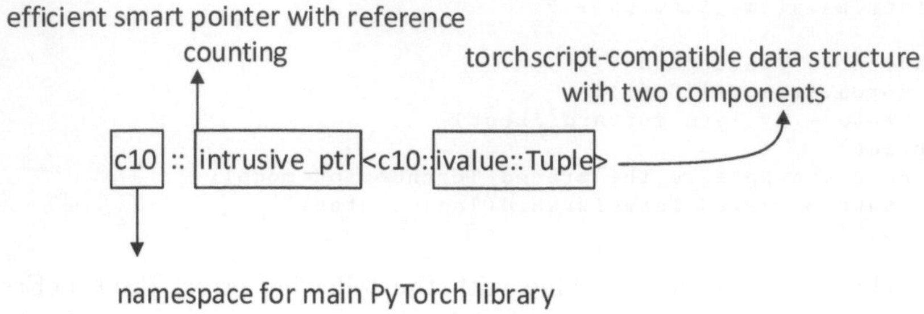

efficient smart pointer with reference
counting

torchscript-compatible data structure
with two components

`c10` `::` `intrusive ptr` `<c10::ivalue::Tuple>`

namespace for main PyTorch library

FIGURE 33.2
Breakdown of the data type used to store LSTM state.

when I was developing the code for this part of the book, I debugged in C++, which was slower and more difficult than Python would have been.

33.4   Using a state-ready TorchScript LSTM model in C++

So, we have a TorchScript LSTM model that can receive a state variable along with its input. We need to learn how to construct and retain the state data in C++. This is not a well-documented part of the libtorch API at the time of writing – I had to carefully experiment to find out the exact requirements for this data structure. Luckily for you, I will now share the fruits of my labour.

33.4.1   The data structure for LSTM state

The first thing to know is that the state tuple is stored in a rather exotic-sounding type: $c10 :: intrusive\_ptr < c10 :: ivalue :: Tuple >$. Let's go through that type and see what it all means. c10 is short for caffe2 – get it? c followed by 2 in binary. Caffe2 is a deep-learning framework that was merged with PyTorch at one point. c10 is a part of the PyTorch library that contains the bulk of the system. Intrusive pointers are smart pointers that manage an object via reference counting. They are more lightweight than the better-known shared pointers as they offload reference counting to the object they wrap instead of implementing it themselves. We have already encountered ivalues, wrappers around torch data structures used with TorchScript models. An ivalue tuple is a version of an ivalue that has two components. In summary, the state data for an LSTM is stored in a smart pointer wrapped around a special tuple-like type of ivalue. You can see a visual representation of this data type in figure 33.2. Add a typedef for this data type to the top of your main.cpp for convenience:

```
//Now I can just use LSTMState to refer to this data type
typedef c10::intrusive_ptr<c10::ivalue::Tuple> LSTMState;
```

33.4.2 Creating initial state

Now you have seen the data type used to represent LSTM state, now referred to as LSTMState, here is some code that will create an initial, random state:

```
LSTMState getRandomStartState(int numLayers, int hiddenSize){
    torch::Tensor h0 = torch::randn({numLayers, hiddenSize});
    torch::Tensor c0 = torch::randn({numLayers, hiddenSize});
    LSTMState state = c10::ivalue::Tuple::create({h0, c0});
    return state;
}
```

I have added parameters to the function to allow for different architectures. The size of the elements in the state tuple is dictated by the number of layers and hidden units per layer. The following Python code illustrates how those parameters are used to create the model. The state data is not concerned with the input size, which is the 'width' of the input signal. For a mono signal, the width is one:

```
input_size = 1 # mono signal
hidden_size = 1
num_layers = 1
seq_length = 5
torch.manual_seed(21)
orig_lstm = torch.nn.LSTM(input_size, hidden_size, num_layers)
input = torch.rand(seq_length, input_size)
h0 = torch.rand(num_layers, hidden_size)
c0 = torch.rand(num_layers, hidden_size)
```

The LSTM networks we will work with only have one layer, but we will experiment with more hidden units, as we did when testing performance.

33.4.3 Passing state into the model

Now, you can create a data structure appropriate for storing the LSTM state and you know how to generate a random start state. This means you are ready to implement state in the WAV processing code. For reference, you can look at section 39.5.22 in the repo guide. The final step is to change the block-based code you had previously so it receives a state. Here is the block processing function:

```
1  LSTMState processBlockState(
2              torch::jit::script::Module& model,
3              const LSTMState& state,
4              std::vector<float>& inBlock,
5              std::vector<float>& outBlock, int numSamples){
6
7     torch::Tensor in_t = torch::from_blob(inBlock.data(), {static_cast<
         int64_t>(numSamples)});
8     in_t = in_t.view({-1, 1});
9     std::vector<torch::jit::IValue> inputs;
10    inputs.push_back(in_t);
11    inputs.push_back(state);
12
13    torch::jit::IValue out_ival = model.forward(inputs);
14    // copy to the outBlock
15    auto out_elements = out_ival.toTuple()->elements();
16    torch::Tensor out_t = out_elements[0].toTensor();
17    out_t = out_t.View({-1});
18    float* data_ptr = out_t.data_ptr<float>();
19    std::copy(data_ptr, data_ptr+inBlock.size(), outBlock.begin());
20    // now retain the state
21    return out_elements[1].toTuple();
22 }
```

The main thing to note in that function is line 11, where we simply add the state value to the ivalue vector after adding the input. TorchScript will unpack that vector into a list of arguments for the forward function. Then, on the last line in the function, we extract the state after processing the input and return it.

When I run this code, the artefacts in the signal at the start of each block are gone. Over to you – can you get this working in your program? Experiment with different network architectures – does it still work with more hidden units? Can you compare performance between the block-based and one-shot versions of the code? Any bottlenecks here?

33.5   Wrap it in a JUCE plugin

The final stage for this chapter is to convert that command-line C++ program into a JUCE plugin. The JUCE plugin will be an effects unit which passes the incoming signal through our simple LSTM model. I should acknowledge that we are implementing this plugin before we even have a good network to process the audio with, but I am keen for you to complete this development cycle now so you have a working plugin for later. In the next chapter, you will learn how to train the network to process the sound to model an amplifier.

So let's get to it! To make the plugin, we will complete the following steps:

1. Create the plugin project and add the necessary variables

2. Load the LSTM model to the plugin

3. Integrate JUCE's processBlock with our LSTM processBlock code

Start with the minimal JUCE and libtorch project from the repo guide section 39.2.11. Make the usual changes to CMakeLists.txt – you want a unique PROD-UCT_NAME and PLUGIN_CODE. Add this header to PluginProcessor.h, and add the typedef:

```
1 #include <torch/script.h>
2 typedef c10::intrusive_ptr<c10::ivalue::Tuple> LSTMState;
```

Add these variables for the LSTM model and its state to the private section of PluginProcessor.h:

```
1 torch::jit::script::Module lstmModel;
2 LSTMState lstmState;
```

Now, set up the plugin's channels to be mono. In PluginProcessor.cpp:

```
1 .withInput  ("Input",  juce::AudioChannelSet::mono(), true)
2 .withOutput ("Output", juce::AudioChannelSet::mono(), true)
```

Compile and run to verify you can build against the TorchScript components.

33.5.1 Load the LSTM model and initialise state

Initialise the model in the constructor in PluginProcessor.cpp. I am using an absolute path for this – change it to match where you have the pt file on your system:

```
1 lstmModel = torch::jit::load("path/to/my_lstm_with_state.pt");
```

Now add the prototype for the getRandomStartState function from the command line program to the private section of your PluginProcessor.h file:

```
1 LSTMState getRandomStartState(int numLayers, int hiddenSize);
```

Add the implementation of getRandomStartState from the main.cpp of repo guide 39.5.22 (seen above) to your PluginProcessor.cpp file. Remember to add the namespace for the class to the function implementation which is AudioPluginAudioProcessor::getRandomStartState. In the constructor of PluginProcessor.cpp, initialise the state after you create the model. So, assuming your model is 1 layer, 1 hidden unit, put this in the constructor:

```
1 lstmState = getRandomStartState(1, 1);
```

Compile and run to verify things are working correctly.

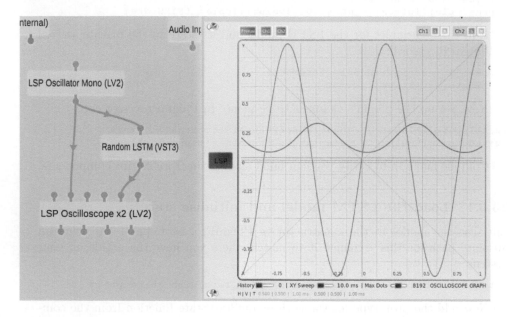

FIGURE 33.3
The LSTM plugin running in the AudioPlugHost test environment, with an oscilloscope showing a sine wave test tone before and after LSTM processing.

33.5.2 Integrate the processBlock code

The final step is to pass incoming audio through the LSTM by integrating the JUCE processBlock function call with the processBlockState function you saw earlier in the command line program. processBlockState has a prototype like this:

```
LSTMState processBlockState(
    torch::jit::script::Module& model,
    const LSTMState& state,
    std::vector<float>& inBlock,
    std::vector<float>& outBlock, int numSamples){
```

JUCE's processBlock function has the following prototype:

```
void processBlock (
        juce::AudioBuffer<float>&,
        juce::MidiBuffer&);
```

So the trick will be converting that juce::AudioBuffer into a format that the processBlockState function can tolerate. The good news is – we already have the two torch-related arguments model and state. We just need an input vector of floats and an output vector of floats. We need to pay some attention to memory assignment here – we do not want to be allocating the memory for buffers of floats in the middle of the audio loop if we can avoid it (noting the limited control we have over the innards of the TorchScript model).

My solution is to create the input and output blocks when prepareToPlay is called – we will know the size of the buffer then. Then, I will use std::copy to copy data from the JUCE AudioBuffer to and from my vector buffers. So – in PluginProcessor.h, add the following vectors to the private section:

```
std::vector<float> inBuffer;
std::vector<float> outBuffer;
```

In PluginProcessor.cpp, in the prepareToPlay function, set them up:

```
inBuffer.resize((size_t)samplesPerBlock);
outBuffer.resize((size_t)samplesPerBlock);
```

Compile and run to check for any mistakes. Now, to get the data from JUCE's processBlock via these vectors into our processBlockState function. In PluginProcessor.cpp, processBlock function, add the following code, which will copy the samples from the incoming block to the inBlock vector:

```
for (int channel = 0; channel < totalNumInputChannels; ++channel)
{
    auto* input = buffer.getReadPointer (channel);
    std::copy(input, input + inBuffer.size(), inBuffer.begin());
    processBlockState(lstmModel, lstmState,
                      inBuffer, outBuffer,
                      buffer.getNumSamples());
    auto* output = buffer.getWritePointer (channel);
```

```
9     std::copy(outBuffer.begin(), outBuffer.begin() + inBuffer.size(),
         output);
10 }
```

Now, compile and run the plugin inside the test host. Try passing a sine signal through it and listening to the result. Figure 33.3 shows the plugin running in the AudioPluginHost test environment. In that figure, I am feeding a test signal through the LSTM and using an oscilloscope plugin from the Linux Studio Plugins project[1] to display the signal before and after processing through the LSTM.

You can make all kinds of extensions to the plugin, but you might want to hold off until you have found out how to train the model to process the sound in a particular way, which you will learn in the next chapter.

33.6   Progress check

At this point you should be able to define an LSTM model in Python, to trace and export it using TorchScript. You can then load the TorchScript model into a C++ program and use it to process audio data. You can also carry out the processing in a block-based model, retaining the state of the LSTM between blocks to avoid artefacts in the processed signal. Finally, you can integrate this functionality into a JUCE plugin.

---

[1]https://lsp-plug.in

34

Training the amp emulator: dataset

In this chapter, I will lead you through the components of the Python training script I have created to train LSTM models to emulate distortion effects and amplifier circuits. The script is based partly on open-source code by Alec Wright, who worked extensively on guitar amplifier modelling with neural networks in 2019–2020. I have rewritten much of Wright's code to align it with current practice in PyTorch programming. My approach in this chapter is to review an existing code base instead of showing you how to create the complete script line-by-line as we did for the C++ code. The main components are the LSTM model, the data loader and the training loop. Along the way, you will find out how to monitor training progress using tensorboard, how to save models for further training later and how to manage training on CPU and GPU devices. At the end of the chapter, I will show you how to import the trained model into a JUCE plugin.

34.1 Training script overview

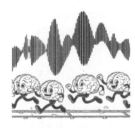

FIGURE 34.1
How fast can LSTMs process audio?

Figure 34.2 shows a high-level overview of training a neural network to emulate a guitar amplifier. You have seen this image in a previous chapter. Let's review the image and see what kind of considerations it raises.

In step 1, you pass a test signal through an amplifier and record the result, which should be straightforward. But what kind of audio signal should you use? How can you ensure you have given the neural network enough examples of how the amplifier processes different types of signals?

In step 2, you pass the signal through the neural network; what kind of neural network model do you need? Is a simple LSTM layer sufficient here? How many hidden units do you need? What about the problem of signal drop-out at the start of a block that we observed in the previous chapter?

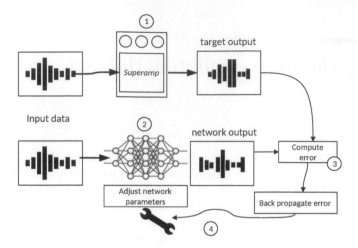

FIGURE 34.2
Four stages to train a neural network. 1: send the test input through the device (e.g. amp) you want to model, 2: send the test input through the neural network, 3: compute the error between the output of the network and amp, 4: update network parameters to reduce error using back-propagation. Back to stage 2.

In step 3, you compute the error between the network output and the amplifier's output. But how will you compare the two output signals? How can you make sure that the network is improving in the right way over time and that it does not learn any strange 'error-avoiding' solutions? Seasoned machine learning engineers reading this will be familiar with the problem of error function exploitation, where the neural network learns to achieve a low error without actually doing what you want it to do! For some entertaining examples of systems exploiting error functions by developing unwanted behaviours, in this case, evolutionary algorithms, I recommend looking at Lehman et al.'s paper[25].

Then, there are stages 4 and 5, where you will use the error to update the network parameters before returning to stage 2. There are various parameters and settings which control the behaviour of the training process, e.g. how much you should adjust the network parameters each time (learning rate), how many inputs the network should process between parameter updates (batch size) and so on.

In the following sections, I will revisit these questions and explain how the training script addresses them.

34.2   Setting up for training

The approach I am taking in this chapter is to work through a Python training program I have created for the book. I will explain the purpose of each part of the program and the decisions made regarding the problems noted above. The first action for you to take is to download the code for the training program and get it working on your machine. The code is described in the repo guide section 39.5.23. To run the code, you will need a virtual environment with the right packages installed. Here is a pip command that will install all the packages you need to run the code:

```
pip install torch torchaudio scipy numpy tensorboard soundfile
    packaging
```

At the time of writing, I am using the versions of those packages shown in the list below.

- torch, Version: 2.1.0

- torchaudio, Version: 2.1.0

- scipy, Version: 1.11.3

- numpy, Version: 1.26.1

- tensorboard, Version: 2.15.0

- soundfile, Version: 0.12.1

- packaging, Version: 23.2

If you install these packages and you have some problems running the script, I recommend that you report these problems as issues on the book's GitHub repository. If you want to install a specific version of a package, you can use a pip command like this:

```
pip install torch==2.1.0
```

The files for the Python training program should be in a 'python' sub-folder in the example code. In that folder, you should see the following files:

1. myk_data.py : functions to prepare data
2. myk_evaluate.py : functions for testing trained networks
3. myk_loss.py : definition of loss functions
4. myk_models.py : definition of models

5. myk_train.py : definition of train and update function

6. train.py : controller program and parameters

You can go ahead and run the train.py script. It will attempt to load a dataset and then train a network. If it cannot find the audio files in the dataset, it should print out some helpful messages and exit. On my system, I see the following output when I run the script – if you see something like the message on line 12: 'epoch, train, val 0 0.7639494 0.7458705 ' you are ready to train!

```
1  python train.py
2  Loading dataset from folder  ../../data/audio_ht1
3  generate_dataset:: Loaded frames from audio file 120
4  Splitting dataset
5  Looking for GPU power
6  cuda device not available/not selected
7  Creating model
8  Creating data loaders
9  Creating optimiser
10 Creating loss functions
11 About to train
12 epoch, train, val  0 0.7639494 0.7458705
```

When you run the script, if you see error messages, read them carefully and fix the problems. Common errors are ModuleNotFound, meaning you have not installed a Python module and assertion errors relating to the location of the audio files. The default script assumes there is a folder two levels up with a data and audio_ht1 sub-folder, as per line 2 in the output above. That is where the training data is located, and you should have received an audio_ht1 folder with the correct data in it when you downloaded the code for the book. I will go into more detail about preparing training data shortly. At this stage, you should be able to run the train.py script and see an output similar to that shown above. Read any errors carefully and work to resolve them before continuing.

34.2.1 Tensorboard

The training program is set up to use tensorboard. Figure 34.3 shows tensorboard in action. It is a machine learning dashboard that lets you observe the progress of training runs using your web browser. If you run the training script, you will see it creates a folder called 'runs'. Each time you run the script, it will create a new sub-folder in the runs folder. Here is an example of the file structure made in the runs folder after running train.py twice:

```
1  |-- Oct24_16-22-50_yogurt52 ht1 LSTM model with 32 hidden units
2  |    |-- events.out.tfevents.1698160970.yogurt52.1869590.0
3  |    '-- saved_models
4  |         |-- 1.wav
5  |         |-- 3.wav
```

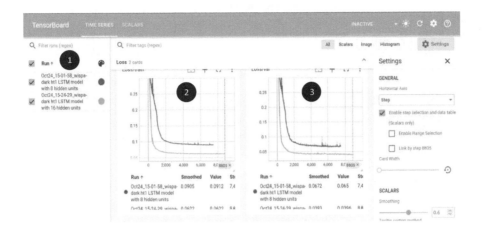

FIGURE 34.3
Tensorboard is a web-based machine learning dashboard. Here, you can see a list
of training runs (1) and graphs showing training progress in terms of training (2)
and validation (3) errors on two separate runs.

```
6  |           |-- lstm_size_32_epoch_1_loss_0.7031.pth
7  |           |-- lstm_size_32_epoch_3_loss_0.6968.pth
8  |           '-- rtneural_model_lstm_32.json
9  '-- Oct24_16-23-03_yogurt52 ht1 LSTM model with 32 hidden units
10     |-- events.out.tfevents.1698160983.yogurt52.1869961.0
11     '-- saved_models
12         |-- 1.wav
13         |-- lstm_size_32_epoch_1_loss_0.756.pth
14         '-- rtneural_model_lstm_32.json
```

If you run tensorboard as follows in the folder with the runs sub-folder in it:

```
1  tensorboard --logdir runs
```

You will see a URL printed to the console, such as http://localhost:6006/. If
you open that URL in your browser, you should see something like the user interface shown in figure 34.3. Try running the train.py script in one terminal window
and then running the tensorboard command in another. Watch the tensorboard
dashboard in your web browser, and you should see the progress of the training
run. I will explain more about the files in the runs folder shortly. Still, as a quick
insight, the pth file is a snapshot of the trained model at a certain point in time,
the JSON file is an exported model in RTNeural format (more on RTNeural later),
and the WAV files are examples of test data being run through the network at the
time the pth file was saved.

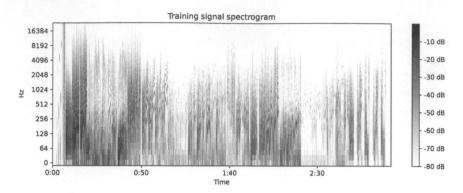

FIGURE 34.4

Spectrogram of the Atkins training signal 'v2_0_0.wav'. You can see the signal is quite varied and dynamic.

34.3 Preparing the dataset

FIGURE 34.5

Capturing training data from a guitar amplifier is similar to 're-amping'.

In broad strokes, the training data consists of recordings of a clean instrument sound and recordings of that instrument sound after it passes through the effects system you wish to model. In practical terms, let's say you want to model a particular guitar overdrive pedal. You would record some clean guitar into your computer. Then, pass that same clean guitar recording through the pedal and record the output as a separate file. So the input and output are the same 'performance', but one is clean and one processed.

You may then wonder how much audio you need and what you need to play on your instrument. Neural Amp Modeller is a neural network-based guitar amp modelling plugin similar to the one we are working on here. According to its author, Steve Atkinson, you will need between three and fifteen minutes of training data. In the Neural Amp Modeller GitHub repository[1], Atkinson provides an audio file that you can use for training called v2_0_0.wav.

The audio file is three minutes long, and it consists of impulses, noise bursts and various unusual guitar-playing sounds, some with pitch

---

[1] https://github.com/sdatkinson/neural-amp-modeler

modulation applied. You can use this file as your training input. You can see a spectrogram of the file in figure 34.4.

As noted above, you then need a recording of that file after it has passed through some sort of signal path. Figure 34.5 illustrates the concept of re-amping. In the example, I am playing the clean training signal out of a sound card and into a small guitar amplifier. Then, I record the output of the amplifier's speaker back into the sound card using a microphone. The recording of the mic would be the training output signal. In practice, I achieved more satisfying results when I recorded directly from the output jack of the amplifier instead of via a mic. Those familiar with sound engineering will know this as 'DI'ing' or direct input. DI'd signal paths are easier to model than fully re-amped signals, as DI'd signals only contain the pre-amplifier and tone circuit. Fully re-amped and mic'd signals include the pre-amplifier circuit, the tone circuit, the power amplifier circuit, the speaker itself and the mic, including the additional reflected signal from the room. You could model such a signal path, but as for more traditional models of modelling guitar signal paths, you might need a more complex model to achieve good results.

If you use longer recordings for your training data, the model will take longer to train, but you might obtain a more thorough model of the effects signal path you are capturing. The thoroughness of the model depends not just on the amount of training data but also on the variety of timbres in the training signal. The training process needs to see what happens to various signals when they pass through your effects chain. Training data with a wide range of timbres will allow the trainer to learn a more dynamic model since the training data will show how the effects respond to different signals.

34.3.1 Sample rate and bit depth

When you prepare your training data audio files, you must decide on the sample rate and bit depth. train.py uses a function called generate_dataset to load the audio files and convert them into the format for training. This function takes a sample rate argument, and it will convert all files to that sample rate. The default is 44,100Hz, but you can use any sample rate you like, so prepare your files with the sample rate you plan to use for training to prevent unnecessary resampling. The samples read from the file will be converted to 32-bit floats, so you can use the standard 16-bit or 24-bit formats. The files should be mono, as the network architecture is designed for mono signals.

34.3.2 Creating training data with digital effects

When you are experimenting with training, you might find it easier to generate some training data by passing the test signal through the plugins you have avail-

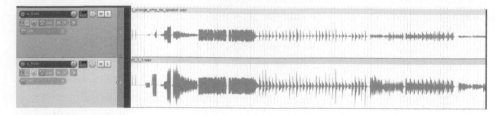

FIGURE 34.6
Clean signal (top) and re-amped signal (bottom) in Reaper.

able to you. For example, you could pass the signal through an amp modelling plugin and see if you can replicate it with the LSTM. Of course, you will be modelling something that might not be very realistic. It is better (and more fun) to pass the signal through an actual, physical pedal or amplifier.

34.3.3 Latency

When you pass a signal out of your computer and then capture it back in via an effects chain, latency is added to the signal due to the buffer used in the audio hardware. Depending on how you capture the signal, the latency may or may not be an issue. If you play the signal from a DAW and record it back to the DAW via the effects you are modelling, you will probably be okay. This is because the DAW will automatically apply latency compensation based on the buffer size of the audio device. If you use some other means to carry out the playback and capture, then you may need to align the captured signal with the clean signal. If you are using some external digital effects in your effects chain, these will add more latency on top of that added by the sound card input and output. Your DAW will not be able to compensate automatically for latency caused by external digital effects. Luckily, the test file 'v2_0_0.wav' has some very short impulses at the start, allowing you to align the recording with the original more easily. If you do not align the files, the training might not work correctly.

34.3.4 What kind of effects can you model?

An important question to address at this point is – what kind of processes and effects can you model with the networks available in train.py? Remember that the LSTM is similar conceptually to IIR filters, and therefore, it works well with the kind of effects that IIR filters can model, such as bandpass filters and EQs. It is also effective at modelling distortion. To model reverbs and other longer-term effects, people tend to use convolutional neural networks. For this chapter, we will focus on LSTM models as it is easy to understand and test them, and they will run in real-time in a plugin format.

34.4 Progress check

At this stage, you should be able to run the training script train.py without errors. You should have conducted some training runs and observed their progress using tensorboard. You may also have created your own training data by passing the provided test signal through an effects unit.

35

Data shapes, LSTM models and loss functions

In this chapter, I will explain in detail the shape of the data and how it is organised for training. My experience has been that figuring out exactly what form the data structure takes can help a lot in understanding what the neural network and training scripts are doing. After that, I will explain why you need to add additional layers to the LSTM model so the 'multichannel' output of the LSTM can be mixed down to a single channel. We will finish the chapter with an examination of the loss function designed to guide the training script to adjust the parameters of the neural network correctly to achieve a usable trained model.

35.1   Sequence length

In the previous chapter, you learned how to set up your training environment and prepare your training data files. Now, you will learn how to use PyTorch's dataset and dataloader functionality to organise that data into a format that can be used for training.

Let's start with a look at the function calls that create the dataset. In train.py, you will find something like the following line:

```
1  dataset = myk_data.generate_dataset(audio_folder + "/input/",
2                                       audio_folder + "/output/",
3                                       frag_len_seconds=0.5)
```

The generate_dataset function receives three arguments:

1. The input data folder. This should contain one or more WAVs. These are the clean audio signals. Essentially, you can just put the test signal WAV file in there.

2. The output data folder. This should contain the processed version of the input WAVs. As for the input, one file will be enough for our purposes.

3. The frag or sequence length, which is the length of the sequence sent into the LSTM each time we call 'forward' during training.

The input and output folder arguments are pretty straightforward, but what exactly does the sequence length do here? Regarding the dataset preparation, the audio files are concatenated into two long sequences of samples, one for the input and one for the output. Then, they are chopped into sub-sequences with the length you specify in the sequence length parameter.

For example, imagine you have 180 seconds of audio at 44,100Hz in the input folder with a matching, processed signal in the output folder. If you ask for a sequence length of 0.5s, you will receive 360 sequences with a length of 22,050 for the input and the same for the output. The sequence length you choose then has an effect on the training process. The longer the sequence, the more information the trainer will have about how the system you are modelling behaves over time.

To put that into more precise terms, the purpose of the LSTM model is to produce the correct output sample for the given input sample. If you have a sequence length of one sample, when you are training, the network only gets to see one sample at a time, and it has to guess what the output should be. This might be okay for a waveshaper-type effect – waveshapers take one input and produce one output; they do not know about any older inputs. But what about some of the delay and filter effects we have seen in previous chapters? They often take account of several previous inputs. Then consider modelling complex analogue circuits such as valve amplifiers – valves are quite 'stateful', which is partly what provides their lively tone. As is often the way with neural networks, experimentation can lead you to the best setting. A good starting point for sequence length is 0.5s, meaning the neural network sees 0.5s worth of input samples and has to predict the following output sample.

35.1.1 Block size

Before we move on, let's return to an important plugin concept: block size. The block size is the number of samples processed in each call to processBlock when a plugin is running. How does the sequence length interact with the block size when the LSTM model is running in a plugin? For example, assuming a sample rate of 44,100Hz, your sequence length during training might be 22,050 samples. Then, when you ultimately run the network in a plugin, it will have a block size of, say, 1024 samples. If we've trained the network with 22,050 previous samples, but the block size is 1024 samples, what is going on here?

The trick is remembering that in plugin mode, we retain the LSTM state between each call to processBlock. This means that an LSTM running in a plugin has access via the LSTM's internal memory cells to some representation of the previous sequence going back to when the plugin started! Way more than 0.5s, potentially. This is the unpredictability of feedback in action. Fortunately, during training, the LSTM learns to focus on that 0.5s of previous audio because it does

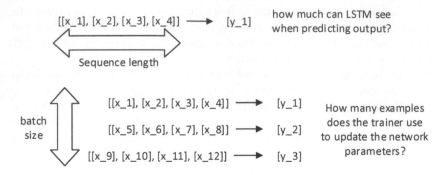

FIGURE 35.1
Sequence length and batch size.

not receive anything else. So, when the network runs in a plugin, it will probably ignore anything older than 0.5s.

35.2   Batch size

While discussing training data shapes and sizes, let us consider the batch size. Elsewhere in the train.py script, we define a variable called batch_size, which is how many sequences will be processed between updates to the network parameters. Updating the network parameters is what training is all about – adjust the network until it processes the signal how we want it to.

What is the difference between batch size and sequence length? The sequence length dictates how much the *network* sees of the previous signal when predicting the following output. The batch size dictates how much the *trainer* knows about the behaviour of the network with different examples of input signals. Figure 35.1 illustrates the difference between sequence length and batch size. A batch consists of several input-output sequence pairs, and after processing a batch, the network's parameters will be adjusted. So, a small batch size means the trainer does not see many examples between updates. So, it does not get a very detailed picture of the error surface it is exploring. Larger batches give more information about the error surface, but that might make it harder to improve because the training 'signal' is more complex.

Luckily for us, there are some very detailed research papers which have explored the ideal sequence length and batch sizes for the problem of guitar amplifier modelling, and those settings provide a good starting point. The defaults for the script are a sequence length of 0.5s and a batch size of 50.

35.2.1 Torch datasets

The generate_dataset function returns a TensorDataset object. It creates that object with the following lines in myk_data.py:

```
1  input_tensor = torch.tensor(np.array(input_fragments))
2  output_tensor = torch.tensor(np.array(output_fragments))
3  dataset = TensorDataset(input_tensor, output_tensor)
```

There are a couple of things to note: 1) we convert the input and output fragments into numpy arrays, and 2) we convert the numpy arrays into torch tensors. The first step is optional, as torch tensors can be created from regular Python lists. But numpy arrays are a more sophisticated data structure than a basic list, allowing for type specification. It is crucial to have control over the data types when you are dealing with large datasets. The following code illustrates how you can specify the data type in numpy:

```
1  import numpy as np
2  data = [[[1], [2], [3]],[[4], [5], [6]]]
3  data_np = np.array(data, dtype=np.float32)
```

Numpy arrays must also have a regular shape. Consider the following code, which will crash:

```
1  import numpy as np
2  data = [[[1], [2], [3]],[[4], [5]]]
3  data_np = np.array(data)
```

Compare it to this code, which will not crash:

```
1  import numpy as np
2  data = [[[1], [2], [3]],[[4], [5], [6]]]
3  data_np = np.array(data)
```

The conversion to a tensor is necessary because tensors are the data structures that torch will accept, as discussed in an earlier chapter.

35.2.2 Dataset shape

The shape of the input_fragments object is (num of sequences, sequence length, sequence width (1 for mono)). So the first dimension indexes through the available sequences, the second dimension indexes through the samples in a given sequence and the final dimension indexes through the channels in the audio. The following diagram illustrates the shapes of the input and output tensors:

```
[                                    [
  [[x_1], [x_2], [x_3], [x_4]],        [[y_1], [y_2], [y_3], [y_4]],
  [[x_5], [x_6], [x_7], [x_8]],        [[y_5], [y_6], [y_7], [y_8]],
  [[x_9], [x_10], [x_11], [x_12]]      [[y_9], [y_10], [y_11], [y_12]]
]                                    ]
```

input tensor   output tensor

To give you some hands-on experience with the shape of the data and how to select things from it, here is some Python code that takes a simple sequence from 0-11 in a 1D array and reshapes it:

```
1  import numpy as np
2  total_len = 12
3  a = np.arange(0, total_len, 1)
4  print(a)
5  channels = 1 # mono
6  seq_len = 2 # 6 samples in a sequence
7  num_seqs = int(len(a) / seq_len / channels )
8  b =  b = a.reshape((num_seqs, seq_len, channels))
9  print("1st seq:", b[0])
10 print("3rd seq:", b[2])
11 # first seq, first sample, first channel
12 print(b[0][0][0])
13 # third seq, second sample, first channel
14 print(b[2][1][0])
```

I see the following output from this code:

```
1  [ 0  1  2  3  4  5  6  7  8  9 10 11]
2  1st seq: [[0] [1]]
3  3rd seq: [[4] [5]]
4  0
5  5
```

In the above, a batch would consist of multiple sequences, e.g. b[0:2]. Selecting within lists is a powerful feature in the Python language. b[0:4] selects indexes 0,1,2 and 3 from b. b[:,1] selects all 'rows' from b, then selects the second column from each row. Experiment with Python's list selection syntax yourself.

35.3   Training, validation and test data

Now you have an understanding of the structure of the complete training data and the interaction between sequence length and batch size, it is time to break the training data into training, test and validation data. Splitting the dataset into training, validation and test data is done with the following line in the train.py script:

```
train_ds, val_ds, test_ds = myk_data.get_train_valid_test_datasets(
    dataset)
```

The get_train_valid_test_datasets function has a 'splits' parameter, which dictates how the data is split into three subsets, and it defaults to splits=[0.8, 0.1, 0.1]. This means the training data is 80%, and the validation and test data are 10% each. The purpose of these three subsets of data is as follows:

- Training data: fed through the network during training to find errors. Errors are fed back to the network to update parameters.

- Validation data: fed through the network during training to check how training is progressing. Validation data errors are not fed back into the adjustment of parameters, so it is considered 'unseen' data. Therefore, it tests how the network performs with data it has not learned from.

- Test data: used to test the network on unseen data at the end of training. This differs from validation data as it is used to compare across training runs, where the training parameters (sequence length, batch size, learning rate) may have been adjusted.

35.3.1   Over-fitting and generalisation

Why do we need to keep 'unseen' data like this? The reason is to avoid over-fitting and to ensure generalisation. Over-fitting is when the network learns to perform very well with the training data, but it does not perform well with any data it has not seen. Imagine a guitar amp that works perfectly if you play a particular riff on a specific guitar through it, but it just generates white noise for any other input. That guitar amp is extremely over-fitted to that particular riff and guitar. This is an exaggeration to make a point, but it gives you the flavour of the over-fitting problem.

Generalisation is when the neural network performs well with data it has not seen. For our imaginary guitar amp, this means it carries out the correct processing regardless of what riff or guitar you play. That is, of course, what you need in a guitar amp. By observing the training progress in terms of performance on the validation data, we are observing how well the network generalises. If during training, the training data performance is getting better, but the validation data performance is getting worse, the network is over-fitting the training data and not generalising well. That is when you should stop training.

35.3.2 Validation data vs. test data

So, that clears up the training and validation data, but what about test data? They both are examples of unseen data. Sometimes, there is an additional process used in training called 'Hyperparameter tuning'. Hyperparameter tuning is different from parameter tuning. Parameter tuning is what the training is doing – it is adjusting the parameters of the neural network, which control how it processes data. Hyperparameter tuning involves adjusting the training parameters, such as batch size, sequence length, learning rate and so on, such that the training process achieves a better score against the validation data.

Since validation data is used to evaluate training in a real sense, we use information from the validation data to adjust the training hyperparameters. Thus, validation data is not 'unseen' between training runs. You might end up with hyperparameter settings that only work for that validation data. That is where the test data comes in – it is a test to use across training runs, whereas validation data is a test to use within a single training run.

35.4 Devices, DataLoaders and Generators

Now we have our data split into training, test and validation data, we are going to prepare it for training. This involves figuring out which compute device we will use and wrapping the data in a PyTorch DataLoader object. Here is the code that train.py uses to select the compute device:

```
1  if not torch.cuda.is_available():
2      device = 'cpu'
3  else:
4      torch.cuda.set_device(0)
5      device = 'cuda'
```

You can see that it either selects the cpu device or the cuda device. CUDA stands for Compute Unified Device Architecture, and it is a framework that PyTorch uses to carry out computations on Nvidia graphics accelerators. It should have been installed on your machine when you installed PyTorch. It will only be available if you have an appropriate type of GPU (basically a recent Nvidia one)

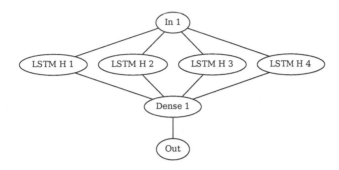

FIGURE 35.2
An LSTM network with a four hidden unit LSTM layer and a densely connected
unit which 'mixes down' the signal to a single channel.

and the correct drivers. You can run this training script on a CPU, but it will take
longer to complete training.

Once you have the device, the next step is to prepare the data loaders. The
data loaders present an interface on the dataset that allows batch processing,
shuffling and other features. It is easy to convert a TensorDataset object into a
DataLoader:

```
1  train_dl = DataLoader(train_ds, batch_size=batch_size, shuffle=True)
```

35.5 Defining the model

Now you have your data in order, you are ready to create your neural network
model. In the previous chapters involving LSTM networks, we used a straightforward model with only an LSTM layer and nothing else. The problem with that
model is the number of output values it generates. For a mono signal, you want
a mono output. But an LSTM layer generates the same width of output as it has
hidden units:

```
1  import torch
2  # LSTM with 1 input, 4 hidden units, 1 layer:
3  lstm = torch.nn.LSTM(1, 4, 1)
4  # send it
5  input = torch.zeros(1, 1)
```

```
6 print(input.shape)
7 output, _ = lstm.forward(input)
8 print(output.shape)
```

This outputs the following, meaning the input is 1x1, but the output is 1x4:

```
1 torch.Size([1, 1])
2 torch.Size([1, 4])
```

To invoke a music technology metaphor, we need to 'mix down' the four channels coming out of the LTSM to a single channel. We can do this by adding a dense layer, creating the network architecture shown in figure 35.2.

```
1 import torch
2 # LSTM with 1 input, 4 hidden units, 1 layer. 1->4
3 lstm = torch.nn.LSTM(1, 4, 1)
4 # Dense layer (linear in torch speak) 4->1
5 dense = torch.nn.Linear(4, 1)
6 # push signal through the LSTM layer
7 output, _ = lstm.forward(torch.zeros((1, 1)))
8 # push signal through dense layer
9 output = dense(output)
10 print(output.shape)
```

This model, which is illustrated in figure 35.2, produces a single output channel:

```
1 torch.Size([1, 1])
```

35.5.1 Defining a class for the model

We now need to wrap up the functionality of our neural network model in a format that makes it compatible with PyTorch. To do this, we define a custom class that extends torch.nn.Module. In that class, we implement a constructor that defines the network layers and a forward function that passes data through the layers the way we want. Since the class extends torch.nn.Module, it can be used with other PyTorch functionalities such as TorchScript and the optimiser you will use shortly. Here are the vital parts of the class: the constructor (called '__init__' in Python) and the 'forward' function. The 'self' argument provides a means for non-static functions to access the object's state. It is similar to 'this' in C++:

```
1 class SimpleLSTM(torch.nn.Module): # extend on torch.nn.Module
2     def __init__(self, hidden_size=32):
3         super().__init__() # call the superclass constructor
4         # Batch first means input data is [batch,sequence,feature]
5         self.lstm = torch.nn.LSTM(1, hidden_size, batch_first=True)
6         # mix down from 32 hidden back to 1 output
7         self.dense = torch.nn.Linear(hidden_size, 1)
8
9     def forward(self, torch_in):
10         x, _ = self.lstm(torch_in)
11         return self.dense(x)
```

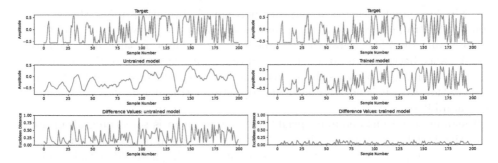

FIGURE 35.3

What does loss mean? The top two plots show extracts from the target output. The middle two plots show the output of an untrained (left) and trained network (right). The right-hand side is much closer to the target. The bottom plots show a simple error between each point in the two plots above. The sum of these values could be a simple loss function.

Some things to note: the LSTM layer is created in batch first mode, meaning the first dimension of the input is the sequence selection index, which is how we've been working. The two layers are assigned to class variables called 'lstm' and 'dense'. The forward function passes the data through the LSTM and then the dense layer. You can test the class as follows:

```
model = SimpleLSTM()
model.forward(torch.zeros((1, 1)))
```

The complete class is defined in the file myk_models.py. If you examine the code, you will see it has some extra functions:zero_on_next_forward and save_for_rtneural. zero_on_next_forward causes a set of zeroes to be passed in for the LSTM state on the next call to forward. It is used to reset the network when necessary during training. save_for_rtneural saves the current model parameters into a JSON file that can be imported into the RTNeural library, allowing for faster inference than is possible with TorchScript. More on that later!

35.6 Loss functions

Now you have the dataset and the model prepared, it is time to look at the design for the loss function. The loss function is responsible for computing the difference between the network's output and the target output in the training data. The loss function needs to capture pertinent information about the signal to guide the training towards the correct settings for the network. Figure 35.3 illustrates an example of a loss that can be easily calculated. In this case, the loss is the distance between the target signal and the signal received from the network. This is an example of perhaps the simplest possible loss function – the Euclidean distance. However, it is not the most appropriate for training models to emulate guitar amplifiers.

Designing loss functions appropriate for your problem is something of an art form. Alec Wright describes a quite complex loss function in his paper about non-linear guitar amp emulation, which we are re-implementing here[45]. Wright's loss function consists of three stages, which he found effective in guiding the training process. Let's consider each of those in turn, but before we do, I will point out that all the loss functions are written using torch functions, as that means they can be computed using torch's parallelisation capabilities where the hardware is available.

35.6.1 Pre-emphasis filter

The first stage in Wright's loss function is a pre-emphasis filter. This is a high-pass filter that removes some of the low-end from the signal. Removing the low end from the signal means the later loss function stages are biased towards better matching in the mid and high frequencies. According to the research paper, Wright et al. found errors in the mid and high frequencies to be a problem during training without the filter. Of course, it is those wonderful high-frequency harmonics that guitarists want to hear. The pre-emphasis filter is implemented in the file myk_loss using a torch convolutional layer. But on examining Wright's code, there are no example coefficients supplied for the high pass filter. The code also creates a convolutional layer with insufficient coefficients to cause high pass filtering. It is more common to use IIR filters, not FIR, for high pass filtering as they require fewer coefficients (as you saw in the preceding chapters about DSP), but the code uses an FIR filter. I also noticed that in the example configurations Wright provides, the pre-emphasis filter is not enabled. In my test runs with my re-working of Wright's code, I was able to achieve good performance (sonically speaking) without the pre-emphasis filter[1]. So, we can ignore the pre-emphasis filter.

35.6.2 Error to signal loss

The next stage after the pre-emphasis filter is the error-to-signal ratio (ESR). The ESR is similar to the distance metric seen in figure 35.3, but it factors in the magnitude of the signal. This means that small losses on quiet signals are as significant

---

[1]https://github.com/Alec-Wright/Automated-GuitarAmpModelling/tree/main/Configs

as larger losses on loud signals, preventing the training from over-focusing on the louder parts of the signal. The following code shows the implementation of ESR loss:

```
def forward(self, output, target):
    self.epsilon = 0.00001
    loss = torch.add(target, -output)
    loss = torch.pow(loss, 2)
    loss = torch.mean(loss)
    energy = torch.mean(torch.pow(target, 2)) + self.epsilon
    loss = torch.div(loss, energy)
    return loss
```

The implementation is located in myk_loss.py. You can use the following code to experiment with the ESR loss. I recommend that you run IPython from inside the 'python' folder in project 39.5.23 from the repo guide:

```
import myk_loss
import torch
esr = myk_loss.ESRLoss()
target = torch.randn((10, 1))
output = target * 0.1 # quite different
print(esr.forward(output, target))
output = target * 0.99 # much closer
print(esr.forward(output, target))
```

Experiment with some different signals to see how the ESR loss comes out.

35.6.3 DC offset loss

The final stage in the loss function is a DC offset loss. This punishes the network if it generates a DC offset, where the output signal appears to be shifted up or down. From our earlier experiments with LSTMs, DC offsets are pretty common. If you have worked with recording technology, you will know that DC offsets are generally undesirable. Here is some code that allows you to experiment with the DC offset loss:

```
def forward(self, output, target):
    loss = torch.pow(torch.add(torch.mean(target, 0), -torch.mean(
        output, 0)), 2)
    loss = torch.mean(loss)
    energy = torch.mean(torch.pow(target, 2)) + self.epsilon
    loss = torch.div(loss, energy)
    return loss
```

Here is some code to experiment with DC offset loss. It generates a signal and then adds different types of DC offset:

```
import myk_loss
import torch
dc = myk_loss.DCLoss()
```

```
 4 target = torch.randn((10, 1))
 5 output = target + 0.1
 6 print(dc.forward(output, target)) # up
 7 output = target - 0.1
 8 print(dc.forward(output, target)) # down
 9 output = target + 0.5
10 print(dc.forward(output, target)) # up a lot
```

The loss function has weightings for each stage, which dictate how much that loss stage inputs into the final loss. By default, ESR loss is weighted at 75% and DC offset at 25%.

35.7   Progress check

At this stage, you should understand the difference between training, validation and test data and their connection to over-fitting and generalisation. You should have experimented with dataset shapes and be able to select content from tensor data structures in different ways. You should create some LSTM models using the SimpleLSTM class and be able to say why the dense layer is necessary. You should also have examined ESR and DC offset loss functions and have experimented by passing signals through those loss functions.

36

The LSTM training loop

In this chapter, I will take you through the LSTM training loop. The training loop is the block of code that carries out the actual training of the neural network, where its parameters are gradually adjusted until it processes the signal correctly. This training loop has some interesting features that work around some of the problems encountered by researchers working on training LSTM networks, for example, using a warm-up step and the exotic-sounding truncated backpropagation through time.

36.1 Overview of the training loop

At this point, you have seen many of the details of the data preparation, the neural network model and the loss function, but you have yet to see the fine details of the training loop, which puts all of these elements into action. The main training loop code is in train.py in example 39.5.23. After setting up the data and the model, as discussed previously, it prepares for training by setting up a folder to save model snapshots as training progresses. The folder is a sub-folder of the log dir generated automatically by SummaryWriter:

```
1  writer = SummaryWriter(comment='32 node LSTM amp')
2  model_save_dir = writer.get_logdir() + "/saved_models/"
3  os.makedirs(os.path.dirname(model_save_dir), exist_ok=True)
```

The script next goes into the main flow of the training loop, which is shown graphically in figure 36.1. In this diagram, you will see how an epoch breaks the data into batches, computing loss for each batch and updating the network parameters. There is a function called myk_train.train_epoch_interval, which does all the work involved in a training epoch. More on that below.

After the epoch completes, the losses are logged for display in the tensorboard. Then various checks occur: save the model weights if the validation loss is a new record, exit if there has been no new record for too long, and exit if there have been more than 'max_epochs' epochs. If the exit tests do not end the script, it runs another epoch.

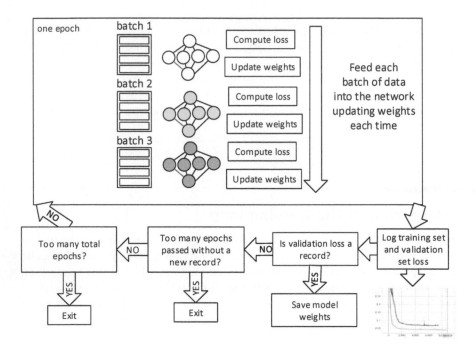

FIGURE 36.1
The training loop. Data is processed in batches with updates to the network parameters between batches. Between epochs, checks are done on whether to save the model and exit.

36.1.1 The optimiser

One component I have yet to mention is the optimiser. It controls how the network parameters are updated. The following lines in train.py set up the optimiser. They also set up a scheduler, which is a kind of controller for the optimiser that varies its settings as the training proceeds:

```
optimiser = torch.optim.Adam(model.parameters(),
                             lr=learning_rate,
                             weight_decay=1e-4)
scheduler = torch.optim.lr_scheduler.ReduceLROnPlateau(
                                        optimiser,
                                        'min',
                                        factor=0.5,
                                        patience=5,
                                        verbose=True)
```

The optimiser's job is to update the network parameters according to the loss. The loss is converted into a set of gradients with respect to the network's parameters using backpropagation (loss.backward()). Intuitively, the gradients show each parameter's influence on the error, so if you follow the gradient by adjusting the parameters, you will reduce the error. The learning rate dictates how far the parameter will be adjusted along that gradient. If the learning rate is too high, the adjustment will be too much, and you will overshoot the ideal setting for the parameter. A high learning rate leads to the loss swinging up and down, never settling on the optimum. If the learning rate is too low, optimisation proceeds too slowly.

This is where the scheduler comes in – it automatically adjusts the learning rate to stop the oscillatory behaviour once you are close to the optimum setting for a parameter.

36.1.2 The train_epoch_interval function

As noted above, the train_epoch_interval function, located in the myk_train module, does much of the work in training for one epoch. It has some interesting features; for example, it carries out a warm-up step. This involves passing the sequence's first section to the neural network and throwing away the loss. This allows the network to warm up, avoiding the dip in the signal you observed before when we investigated the effect of state in LSTMs (figure 33.1). The default warm-up length is 1000 samples, which leaves 21,050 samples for the actual loss calculation. Another feature is that it uses the model's ability to zero out its state via its zero_on_next_forward function. It does this between batches to ensure the LSTM's memory is clear.

Another feature of this epoch training function is the idea of interval training. This is where the network parameters are updated after only part of a sequence has gone through, e.g. every 2046 samples. The state of the network is retained as this proceeds. This technique has the fancy name of Truncated backpropagation through time. It is used because the shorter sequence lengths allow for more frequent updates to the parameters with less computational complexity than for longer sequences, which, according to Wright, leads to better training[46].

The train_epoch_interval function ultimately returns the mean loss across all batches. The main loop then logs this training loss as well as computing and logging the validation set loss.

36.2   Putting it all together: training an LSTM

Now, you have seen all the critical components in the training process. I will now present the workflow as a series of steps that you can follow to train an LSTM network to model a guitar amp.

1. Load the training signal into your DAW

2. Wire things up so the DAW can play the training signal out to a guitar amplifier, then record the processed signal back in

3. Check the alignment of the training signal and the recording and make sure the impulses at the start of the test signal line up

4. Export the recorded signal in the same sample rate as the training signal

5. Prepare the data set folders: one folder containing the training signal WAV and one containing the processed signal

6. Edit settings in train.py to reflect the location of your WAV files

7. Edit other settings in train.py, especially lstm_hidden_size and expt_desc

8. Activate your Python virtual env and run train.py

9. Run tensorboard with the logdir argument pointing at the folder where you ran train.py and open the tensorboard dashboard in your web browser

10. Wait for training to proceed. Training is complete when the validation loss curve flattens out

11. If you observe high levels of oscillation in the loss curve, reduce the learning rate and start again

36.3 Examples of training runs

I carried out some experimentation and analysis using the train.py training script. I tried training different-sized networks with different distortion effects. I have presented some results from training to emulate a Blackstar HT-1 valve guitar amplifier in figure 36.2. I found that the best architecture was a 32-unit LSTM, which, as you can see in the figure, achieved a very close match to the target output. The 256-unit LSTM also achieved a close match, but not as close, and it uses a lot more computation with many more LSTM units.

36.4 Progress check

That is the end of the information about training. There was a lot of detail there – this is hard-won information that researchers and their computers have sweated over for several years. So, at this point, you should have successfully run the training script. I hope you also created your own dataset using either a digital distortion effect or a real amplifier or distortion circuit. In the following chapters, you will need a trained model to use, so you should pause here and try to train a model for yourself before proceeding.

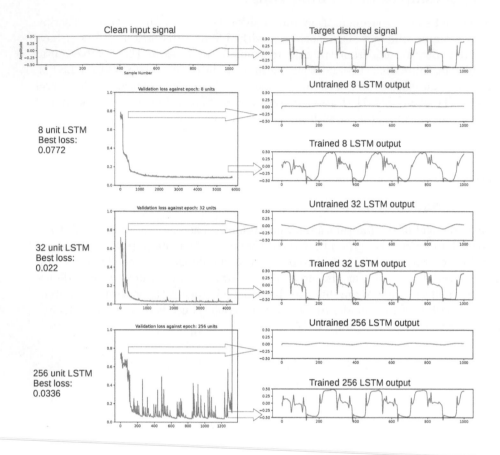

FIGURE 36.2
Comparison of training runs with different sized LSTM networks. At the top you can see the input signal and the target output signal recorded from Blackstar HT-1 valve guitar amplifier. The descending graphs on the left show the validation loss over time for three LSTM network sizes. The waveforms show outputs from the networks before and after training.

37

Operationalising the model in a plugin

In this chapter, I will show you how to operationalise your trained LSTM models in a plugin. You have already seen how to run a TorchScript model in a JUCE plugin project in a previous chapter, but here, I will update that code to cope with the final version of the model. This model version is more straightforward because it maintains its state internally instead of relying on you managing it externally. At the end of the chapter, you should have a fully trained LSTM model running inside a plugin in your plugin host.

37.1   Exporting the model using TorchScript

Once you have a trained model, you can load it into the JUCE plugin you created previously and then use the plugin as a real-time effects unit. The first step is to export the trained model into a format that TorchScript can use. During training, snapshots of the model are saved by the training script as '.pth' files every time the model hits a record validation loss. If you want the best model, you need to locate the best scoring .pth file in your 'runs' folder. Luckily, the training script writes the validation loss into the filenames of the .pth files. Here is an example of the files created in a run:

```
|-- Oct24_16-22-50_yogurt52 ht1 LSTM model with 32 hidden units
   |-- events.out.tfevents.1698160970.yogurt52.1869590.0
   '-- saved_models
       |-- 1.wav
       |-- 3.wav
       |-- lstm_size_32_epoch_1_loss_0.7031.pth
       |-- lstm_size_32_epoch_3_loss_0.6968.pth
       '-- rtneural_model_lstm_32.json
```

There is a file called 'lstm_size_32_epoch_3_loss_0.6968.pth'. That is the file you want to convert into a TorchScript model file. The following Python code will load a model from a pth file and export it using TorchScript:

```
import torch
import myk_models # make sure this file is in the same folder
```

315

```
 3
 4 # set this to the name of your actual pth file
 5 # possibly with its full path
 6 saved_pth_path = 'lstm_size_32_epoch_3_loss_0.6968.pth'
 7 export_pt_path = 'dist_32.ts'
 8 # load from pth
 9 model = torch.load(saved_pth_path)
10 model.eval()
11 # save
12 scripted_model = torch.jit.script(model)
13 torch.jit.save(scripted_model, export_pt_path)
```

This is slightly different from how we exported TorchScript models in chapter 32 – instead of tracing the model, we call jit.script. This is necessary because the whole model, with its dense layer and internal handling of the LSTM state, has some logic in its forward block, which a simple trace cannot handle.

The result of running that code is a file called dist_32.ts. I have given it the 'made-up' .ts extension so I can see it is a TorchScript file.

37.2   Set up the JUCE project

Now you have your model, you can create the JUCE project. The JUCE + libtorch starter project described in section 39.2.11 is a good starting point. In CMakeLists.txt, the main things to change are the PLUGIN_CODE and PROD-UCT_NAME properties and ensuring the TORCH path is correct for your system. Build the project to verify things are working correctly. The standalone program should launch and present a message from libtorch. You can find a completed implementation of the following steps in the project described in section 39.5.24 – there are some large blocks of code in this chapter, so please feel free to refer to that complete version when you need to.

Once you have the default torch and JUCE project building, go into Plugin-Processor.h and add the TorchScript header:

```
1 #include <torch/script.h>
```

Then add the following lines to the private section of PluginProcessor.h:

```
1 torch::jit::script::Module lstmModel;
2 std::vector<float> inBuffer;
3 std::vector<float> outBuffer;
4 void processBlockNN(torch::jit::script::Module& model,
5                     std::vector<float>& inBlock,
6                     std::vector<float>& outBlock,
7                     int numSamples);
```

lstmModel will store the model loaded with TorchScript. You have seen the inBuffer and outBuffer idea before when we were working with the simple, random LSTM model – they are used to pass the audio data from the JUCE processBlock function to the processBlockNN function. The processBlockNN will pass the data through the model.

Now, over to PluginProcessor.cpp in the constructor. Set the audio channels to mono:

```
.withInput ("Input",   juce::AudioChannelSet::mono(),
...
.withOutput ("Output", juce::AudioChannelSet::mono(),
...
```

Then, in the constructor, load the model. We shall hard code the path to the model for now. In the following example, I am loading from C:\temp\models on Windows, and I have added some code to double-check that the file exists and to crash if not:

```
// might need this at the top of the file
#include <filesystem>
...
std::string modelFolder {"C:\\temp\\models\\"};
std::string fp {modelFolder + "dist_32.ts"};
if (!std::filesystem::exists(fp)){
    DBG("File " << fp << "not found");
    throw std::exception();
}
DBG("Loading model from " << fp);
lstmModel = torch::jit::load(fp);
```

Critical note for Windows users: remember you must build in the same mode as the version of libtorch you are using, e.g. Release or Debug mode, or your program will silently and un-debuggably crash on the jit load statement.

37.2.1   processBlock via neural network implementation

Before attempting a build, put in the following implementation of the process-BlockNN function to PluginProcessor.cpp:

```
void AudioPluginAudioProcessor::processBlockNN
        (torch::jit::script::Module& model,
         std::vector<float>& inBlock,
         std::vector<float>& outBlock,
         int numSamples){
  // convert input block to tensor
  torch::Tensor in_t = torch::from_blob(inBlock.data(),
                                        {static_cast<int64_t>(
    numSamples)});
  //Change tensor shape to 3D
  in_t = in_t.View({1, -1, 1});
  // prepare inputs as IValue vector
  std::vector<torch::jit::IValue> inputs;
  inputs.push_back(in_t);
  // process inputs
  torch::jit::IValue out_ival = model.forward(inputs);
  // extract outputs
  torch::Tensor out_t = out_ival.toTensor();
  out_t = out_t.View({-1});
  // copy outputs to output block
  float* data_ptr = out_t.data_ptr<float>();
  std::copy(data_ptr, data_ptr+inBlock.size(), outBlock.begin());
}
```

This is very similar to the code you saw earlier when working with the random LSTM networks in JUCE. The difference is that we are no longer managing the state as the state is now internally managed by the model, and the input tensor is 3D instead of 2D. To make the 3D tensor from the 1D list from the signal buffer, we call view(1, -1, 1) instead of view(-1, 1) to reshape the tensor. The 3D tensor is needed as the model class SimpleLSTM has been set up to process batches of sequences instead of single sequences. So, we just create a batch containing one sequence. Now, build the project and verify that execution gets past the call to torch::jit::load in the constructor. As mentioned above, you must build in release mode on Windows unless using a debug build of libtorch.

37.2.2   Passing live audio data to processBlockNN

The last step in operationalising the trained LSTM model inside a JUCE plugin is to transfer the audio data received in the plugin's processBlock function over to the processBlockNN function to pass it through the neural network. First, you need to set up the buffers which will transport that data. In PluginProcessor.cpp's prepareToPlay function, set up those buffers you added earlier:

```
inBuffer.resize((size_t)samplesPerBlock);
```

```
2  outBuffer.resize((size_t)samplesPerBlock);
```

In PluginProcessor.cpp's processBlock function, fill up the buffers as follows:

```
1  for (int channel = 0; channel < totalNumInputChannels; ++channel)
2  {
3      // so you can read from the incoming audio buffer
4      auto* input = buffer.getReadPointer (channel);
5      // copy from incoming audio to inBuffer
6      std::copy(input, input + inBuffer.size(), inBuffer.begin());
7      // pass it through the network
8      processBlockNN(lstmModel, inBuffer, outBuffer, buffer.
       getNumSamples());
9      // so you can write to the outgoing audio buffer
10     auto* output = buffer.getWritePointer (channel);
11     // copy to the outgoing audio buffer
12     std::copy(outBuffer.begin(), outBuffer.begin() + inBuffer.size(),
        output);
13 }
```

That does the work of ferrying the audio to the neural network and back again. Compile and test. If you are lucky or running on Linux, you might hear the neural network working its magic on your signal and emulating whatever you trained it to emulate. But it is more likely that you will hit the performance limit of TorchScript, and you will hear the audio glitching. For example, with a 32-unit LSTM layer running on my M1 Mac Mini (2020 model), the plugin runs ok in the AudioPluginHost, but in Reaper, I can hear each block being processed, followed by a short silence. If I render the audio from Reaper, it renders at 0.2x real-time, but the result sounds correct. The plugin runs surprisingly well on my Linux Intel 10th Gen i7 machine. This shows that the best-case scenario performance figures we looked at in section sec:lstm-performance are far from reality.

37.3   Progress check

At this point, you should have a working JUCE plugin that loads in a trained LSTM model using TorchScript. You might find that the model you load will not run in a plugin host without audible dropouts in the signal. Experiment with different sizes of LSTM layers. See if you can train networks with varying numbers of units or export differently sized, untrained networks. How many units can your machine/ DAW combination run in real-time without audio dropouts?

38

Faster LSTM using RTNeural

In this chapter, you will learn how to make your neural network process audio faster using the RTNeural inferencing engine. You will find out why RTNeural exists and who created it. Then, you will learn how to export your model's weights to a JSON file, which can be read back in and used to set up an RTNeural network. You will see a comparison of RTNeural and TorchScript, showing that models with the same weights and architecture in both systems output the same values. You will see that RTNeural models run two or three times faster than TorchScript models. Finally, you will see how you can deploy RTNeural in a JUCE plugin, which is the conclusion of our technical work with neural effects.

38.1 What is RTNeural?

If you experimented with different-sized LSTMs in the previous section, you might have concluded that TorchScript needs to be faster for your needs. Luckily, RTNeural exists, and it can run neural networks faster than TorchScript. RTNeural is a neural network inferencing library written in C++ by Jatin Chowdhury and released in 2021[6]. RTNeural implements some common types of neural network components, such as LSTM units, convolutional units, and so on, in efficient C++ code. The implementation is efficient but also visible to the compiler, allowing it to further optimise. RTNeural can only perform inferencing (the forward function) – it does not implement training capabilities. This means you need to train in PyTorch (or tensorflow) and export the trained model's weights in a special JSON format that RTNeural can understand.

38.2   Exporting model weights

Before you dive in and swap out TorchScript for RTNeural in your plugin, let's start with a command-line test program that demonstrates how to create an RT-Neural 'clone' of a torch model. The code for this example is described in section 39.5.25. Open up that project in your IDE. You will find a Python folder and a CMake / C++ project. In the Python folder, there is a script called export.py that is very similar to the one found at the start of this chapter in section 37.1. In fact, the only additional line is the call to save_for_rtneural. This call generates a JSON file from the model weights. The JSON file is how we will send the weights over to RTNeural. The JSON file looks like this – essentially a dictionary with several arrays of floating point values:

```
1 {"lstm.weight_ih_l0": [[0.0794372409582138], [0.03175497055053711],
     [0.07416573166847229], [0.0253373384475708
2 ], [-0.06248393654823303], ...
```

That JSON is generated by the myk_models.SimpleLSTM model's function save_for_rtneural, and it has a key for each layer in the network, along with the parameter settings for that layer.

38.3   Defining models

Next, you will find two files in the project's src folder. main_rtneural_basic.cpp and main_torchscript_vs_rtneural.cpp. The first is a minimal example of defining an RTNeural model and then loading network parameters from a JSON file. To define a model, you need some lines like this:

```
1 const int lstm_units = 64;
2 using MyLSTMType = RTNeural::ModelT<float, 1, 1,
3          RTNeural::LSTMLayerT<float, 1, lstm_units>,
4          RTNeural::DenseT<float, lstm_units, 1>>;
```

That is quite a complex type specifier – let's break it apart. This line sets up a new type of model called MyLSTMType. You can call it whatever you want. The new model will process floats and have one input and one output.

```
1 using MyLSTMType = RTNeural::ModelT<float, 1, 1,
```

The following two lines specify the layers: an LSTM layer with one input and 'lstm_units' units followed by a Dense layer with lstm_units inputs and one output.

```
1 RTNeural::LSTMLayerT<float, 1, lstm_units>,
2 RTNeural::DenseT<float, lstm_units, 1>
```

Once you have this type defined, you can create an instance of a model of that type, and then you load weights into it layer by layer:

```
1  // create a model using the new type
2  MyLSTMType model;
3
4  // create a json object from a file
5  std::ifstream jsonStream("lstm_weights.json", std::ifstream::binary);
6  nlohmann::json modelJson;
7  jsonStream >> modelJson;
8
9  // get the lstm layer from the model
10 auto& lstm = model.get<0>();
11 // write the weights from the json object into the layer
12 RTNeural::torch_helpers::loadLSTM<float> (modelJson, "lstm.", lstm);
13
14 // get the dense layer
15 auto& dense = model.get<1>();
16 // load the weights from the JSON object into the layer
17 RTNeural::torch_helpers::loadDense<float> (modelJson, "dense.", dense
      );
```

In the example code in main_rtneural_basic.cpp, you can see this in action. I added some sanity-checking code that verifies that the number of units in the model is the same as the number of units in the exported JSON. The program creates a model, loads in weights, and then passes some numbers through the model.

38.4   RTNeural output validation

The second C++ program included with the project verifies that the output received from the RTNeural model is the same as the output from a TorchScript model exported from the same trained model. If you find the target rtneural-vs-script in the project and build and run it, you should see output something like this, which compares the output received from the RTNeural and the TorchScript versions of the same network architectures with the same weights loaded:

```
1  RT: 0.085324 TS: 0.085324
2  RT: 0.099943 TS: 0.099943
3  RT: 0.104003 TS: 0.104003
4  RT: 0.098274 TS: 0.098274
```

As you can see, the output from both models is identical.

38.5   RTNeural performance

I tested the performance of RTNeural by creating two-layer LSTM → Dense networks with increasing numbers of LSTM units. I passed 44,100 values through the networks multiple times and computed the average time taken. Compared to an equivalent TorchScript network, the RTNeural networks ran two or three times as fast. If you are that way inclined, you can repeat my experiment using the code (performance.cpp) you can find in the project in section 39.5.25. Some performance figures also show similar results in Chowdhury's RTNeural paper[6].

There are some limitations to RTNeural. The main ones are that it only has a subset of the available torch modules, and its performance gain vs TorchScript decreases as the network size increases.

There are other accelerated neural network inference engines available. For example, ONNX. Google's Magenta team have even provided plugins that run the neural network code in an embedded javascript runtime, exploiting the capability of tensorflow.js to run accelerated in a WebGL context. I leave it to the reader to investigate the other options.

38.6   JUCE plugin using RTNeural

To complete our work with RTNeural, you will deploy an RTNeural model inside a JUCE plugin. You can start with a basic JUCE/ CMake plugin project – you do not need libtorch in this project. The project in section 39.2.2 is a good starting point. Open up the project folder and add these two lines to CMakeLists.txt:

```
1 # you should have this line already
2 add_subdirectory(../../JUCE ./JUCE)
3 # add these two lines below it:
4 add_subdirectory(../RTNeural ${CMAKE_CURRENT_BINARY_DIR}/RTNeural)
5 include_directories(../RTNeural)
```

This assumes you have a folder above your project's folder called RTNeural containing a clone of the RTNeural GitHub repository. It should already be there if you are working with the code pack for the book. Next, change the PLUGIN_CODE and PLUGIN_NAME properties in CMakeLists.txt as you see fit. You should also ensure that IS_SYNTH is set to FALSE so your plugin can be an effects processor. You can attempt a test build to verify you have your RTNeural and JUCE paths set correctly.

38.6.1 Define the model architecture

The next step is to define the model architecture – just as we did in the command-line RTNeural program. You can add the specification to the PluginProcessor.h file:

```
1 #include <RTNeural/RTNeural.h>
2 using RTLSTMModel32 =
3           RTNeural::ModelT<float, 1, 1,
4             RTNeural::LSTMLayerT<float, 1, 32>,
5             RTNeural::DenseT<float, 32, 1>>;
```

Notice that I have given the model specification the name 'RTLSTMModel32', which will make it clear (to me later!) that it is a 32-unit model. Next, we can add an object of that type to the private section of PluginProcessor.h:

```
1 RTLSTMModel32 lstmModel;
```

38.6.2 Load weights into the model

Then, let's add the function to load some weights from a JSON file into that model:

```
1  // prototype in PluginProcessor.h:
2  void setupModel(RTLSTMModel32& model,
3                  std::string jsonFile);
4  // implementation in PluginProcessor.cpp:
5  void TestPluginAudioProcessor::setupModel(RTLSTMModel32& model,
6                                            std::string jsonFile){
7      int lstm_units = 32;
8      // read in the json file
9      std::ifstream jsonStream(jsonFile, std::ifstream::binary);
10     nlohmann::json modelJson;
11     jsonStream >> modelJson;
12
13     // get the lstm layer and check that the weight count is correct
14     auto& lstm = model.get<0>();
15     const int json_lstm_size = modelJson["lstm.weight_ih_l0"].size()
       /4;
16     if (json_lstm_size != lstm_units){
17         std::cout << "Model wants " << lstm_units
18                   << " lstm units but JSON file specifies "
19                   << json_lstm_size << std::endl;
20         throw(std::exception()); // crash if not correct
21     }
22     // load in the lstm weights
23     RTNeural::torch_helpers::loadLSTM<float> (modelJson, "lstm.",
       lstm);
24     auto& dense = model.get<1>();
25     RTNeural::torch_helpers::loadDense<float> (modelJson, "dense.",
       dense);
26 }
```

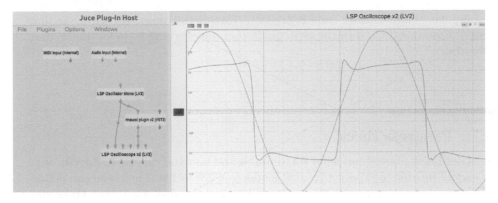

FIGURE 38.1

A sinusoidal test signal passing through an RTNeural LSTM distortion effect in AudioPluginHost. The sinusoidal wave is the original signal, the clipped out wave is the LSTM-processed signal.

Then you can call that code from your PluginProcessor constructor in Plugin-Processor.cpp:

```
setupModel(lstmModel, "path-to-json-file");
```

38.6.3 Send audio through the model

Now tell your plugin to be a mono plugin by editing the constructor in Plugin-Processor.cpp:

```
.withInput("Input",  juce::AudioChannelSet::mono(),
...
.withOutput("Output", juce::AudioChannelSet::mono(),
...
```

Then implement processBlock such that it sends each incoming sample through the LSTM network – in PluginProcessor.cpp:

```
for (int channel = 0; channel < totalNumInputChannels; ++channel){
    auto* outData = buffer.getWritePointer (channel);
    auto* inData = buffer.getReadPointer (channel);
    for (auto i=0; i<buffer.getNumSamples(); ++i){
        // as simple as this!
        outData[i] = lstmModel.forward(&inData[i]);
    }
}
```

At this point, you probably think this is a lot simpler than TorchScript, especially when preparing the data to be sent to the network. You may be correct, but the bespoke weight-loading code is tricky – TorchScript does this with one line of code. Example 39.5.26 in the repo guide contains working code for JUCE and RTNeural.

38.7 Progress check

You should now understand what RTNeural is and how you can use it. Specifically, you should be able to export a trained model to a JSON file. Then, you can define an RTNeural model using the ModelT template and the available layers. Following that, you can load the JSON file into your program and configure the weights in your RTNeural model from the values in the file. You saw how you could compare the output of a TorchScript model to that of an RTNeural model with the same architecture and weights and that the outputs were the same. You completed the chapter by deploying an RTNeural model in a JUCE plugin and using it to process a real-time audio signal.

38.8 Control parameters for neural effects

In this part of the book, we have deeply explored a specific problem, i.e., modelling non-linear guitar amplifier signal paths. One aspect of this problem we did not consider is adding controls to the plugin, such as tone control. An easy way to do this is to add an IIR filter to the signal path before or after the neural network. It will not be an emulation of non-linear tone control, but it will allow the user to adjust the tone and still have the non-linear behaviour of the main effect.

The more proper way to implement a tone control is to train the neural network to know how it should behave with different tone control settings. It is actually easier than you might think to do this, but unfortunately, it is out of scope for this edition of the book to explain it fully here. But I can certainly give you a hint as to how to do it. You must adjust your training data if you want to add a control parameter. Play the test signal through the amplifier three times. For the first time, set the amplifier's tone control to zero, then halfway around, then put it to 10 or 11 if you have that kind of power. For more detailed modelling, you can increase the number of positions. Now you have three training examples with different settings of the tone knob. Then, you need to adjust the training script so that the input signal has the audio samples in one channel and the setting for the tone control in another channel – so each 'sample' in the training input has two values. You will need to adjust the network architecture so it can take two values instead of one at its input. Eventually, you will run the trained model in a plugin. The difference here will be that you need to pass in two values: the audio signal and the control parameter setting. Good luck!

38.9 What other effects can you model with neural networks?

Neural networks can model many other types of effects. I do not have space to cover any more effects types here, but the general principles of training and deployment are undoubtedly transferable. As a bonus, I have provided some starter and experimental code for a reverberator based on Christian Steinmetz's paper[39]. It is described in the code repo section 39.5.27. An area of rapid growth in neural effects is differentiable effects. You can find lots of information about those in Lee et al.'s review paper[24].

38.10 Concluding neural effects ... and the book

Well done – you have reached the end of neural effects. You may have noticed that this was the longest part of the book, something of an epic. I extended it like this because I enjoyed exploring all aspects of DSP and neural signal processing and wanted to share as much information as possible. I hope you enjoyed the journey, too. This is also the end of the book. I have very much enjoyed bringing all of this information out of the research papers and GitHub repositories and presenting it to you. I hope the book sends you into an exciting future of AI-enhanced audio software development.

39

Guide to the projects in the repository

In this chapter, I will tell you where you can find the projects referred to throughout the book in the GitHub source code repository that accompanies the book.

39.1 Complete projects

Firstly, here are the final versions for each of the three examples:

39.1.1 Meta-controller

```
|-- Part2_MetaController
|    '-- 010e_metacontroller
```

39.1.2 Improviser

```
|-- Part3_Improviser
|    |-- 020h_midi_markov_vel
```

39.1.3 Neural FX

```
|-- Part4_NeuralFX
|    |-- 037e_lstm-rtneural-JUCE
```

39.1.4 Test tools

Next, there are a few test tools. These are located in the TestTools folder.

```
|-- TestTools
|    |-- AudioPluginHost
|    |-- MykScope
```

39.2   Part 1: Getting started

39.2.1   001-hello-cmake

```
|-- Part1_GettingStarted
|    |-- 001_hello_cmake
```

39.2.2   002-cmake-juce

```
|-- Part1_GettingStarted
|    |-- 002_minimal_plugin_cmake
```

39.2.3   003x-sineplugin series

```
|-- Part1_GettingStarted
|    |-- 003a_sineplugin
|    |-- 003b_sineplugin_midi
|    |-- 003c_sineplugin_env
```

39.2.4   Challenge solution: 003 sineplugin

```
|-- Part1_GettingStarted
|    |-- 003a_sineplugin
```

39.2.5   Sineplugim with midi control

```
|-- Part1_GettingStarted
|    |-- 003b_sineplugin_midi
```

39.2.6   Sineplugin with envelope generator

```
|-- Part1_GettingStarted
|    |-- 003c_sineplugin_env
```

39.2.7   FM plugin with slider controls

```
|-- Part1_GettingStarted
|    |-- 004a_fmplugin-basic
```

39.2.8   FM plugin with proper parameters

```
|-- Part1_GettingStarted
|    |-- 004b_fmplugin-params
```

39.2.9   FM plugin with parameters and drone mode

```
|-- Part1_GettingStarted
|    |-- 004c_fmplugin-params-
drone
```

39.2.10   Minimal libtorch

```
|-- Part1_GettingStarted
|    |-- 005b_minimal_libtorch
```

39.2.11   Minimal libtorch and JUCE

```
|-- Part1_GettingStarted
|    '-- 008_libtorch_and_juce
```

39.3 Part 2: meta-controller

39.3.1 Linear 2D regression

```
|-- Part2_MetaController
|    |-- 006a_libtorch_regressor
```

39.3.2 Two parameter FM synthesizer

```
|-- Part1_GettingStarted
|    |-- 004c_fmplugin-params-
drone
```

39.3.3 Two parameter FM synthesizer with super-knob

```
|-- Part2_MetaController
|    |-- 007_fmplugin_superknob
```

39.3.4 Two parameter FM synthesizer with torch-knob

```
|-- Part2_MetaController
|    |-- 009
     c_fmplugin_torchknob_train
```

39.3.5 Clean starter project for plugin host

```
|-- Part2_MetaController
|    |-- 010a_plugin_host_starter
```

39.3.6 Super basic plugin host

```
|-- Part2_MetaController
|    |-- 010b_plugin_host_basic
```

39.3.7 Plugin host with graph and load button

```
|-- Part2_MetaController
|    |-- 010c_plugin_host_graph
```

39.4 Part 3: Improviser

39.4.1 Command line Markov model starter project

```
|-- Part3_Improviser
|    |-- 020a_markov-starter
```

39.4.2 JUCE Markov model starter project

```
|-- Part3_Improviser
|    |-- 020b_juce-markov-starter
```

39.4.3 Monophonic MIDI markov effect

```
|-- Part3_Improviser
|    |-- 020c_midi_markov_pitch
```

39.4.4 Monophonic MIDI markov with inter-onset-intervals

```
|-- Part3_Improviser
|    |-- 020d_midi_markov_ioi
```

39.4.5 Monophonic MIDI markov with IOI and duration

```
|-- Part3_Improviser
|    |-- 020e_midi_markov_duration
```

39.4.6 ChordDetector class

```
|-- Part3_Improviser
|    |-- 020f_chord_detector
```

39.4.7 Polyphonic MIDI markov

```
|-- Part3_Improviser
|    |-- 020g_midi_markov_poly
```

39.4.8 Polyphonic MIDI markov with velocity

```
|-- Part3_Improviser
|    |-- 020h_midi_markov_vel
```

39.5 Part 4: Neural FX

39.5.1 Reading and writing WAVs with tinywav

```
|-- Part4_NeuralFX
|    |-- 029a_three_pole
```

39.5.2 Tinywav starter code

```
|-- Part4_NeuralFX
|    |-- 031a_tinywav
```

39.5.3 Time domain convolution: moving average

```
|-- Part4_NeuralFX
|    |-- 031b_convolution_time-
domain
```

39.5.4 Plotting spectrums of WAV files

See utils/two_wav_spectrums.py

```
|-- Part4_NeuralFX
|    '-- utils
```

39.5.5 Time domain convolution: performance

```
|-- Part4_NeuralFX
|    |-- 031c_convolution_time-
domain-speed
```

39.5.6 Frequency domain convolution

```
|-- Part4_NeuralFX
|    |-- 032c_convolution_freq-
domain
```

39.5.7 JUCE convolution plugin

```
|-- Part4_NeuralFX
|    |-- 033a_convolution_JUCE
```

39.5.8 One-pole IIR basic

```
|-- Part4_NeuralFX
|    |-- 034a_iir_basic
```

39.5.9 Many-pole IIR basic

```
|-- Part4_NeuralFX
|    |-- 034b_iir_general
```

39.5.10 IIR filter design Python script

See utils/iir_filter_design.py

```
|-- Part4_NeuralFX
|    |-- utils
```

39.5.11 IIR filter with JUCE DSP module

```
|-- Part4_NeuralFX
|    |-- 034d_iir_JUCE
```

39.5.12 Command-line wave-shaper

```
|-- Part4_NeuralFX
|    |-- 035a_waveshaper_basic
```

39.5.13 Basic JUCE wave-shaper

```
|-- Part4_NeuralFX
|    |-- 035b_waveshaper_JUCE
```

39.5.14 JUCE waveshaper with DSP modules

```
|-- Part4_NeuralFX
|    |-- 035c_waveshaper_JUCE_DSP
```

39.5.15 Amp emulator using classical DSP techniques

035d_JUCE_processor_chain

```
|-- Part4_NeuralFX
|    |-- 035d_JUCE_processor_chain
```

39.5.16 Basic Python LSTM examples

```
|-- Part4_NeuralFX
|    |-- 036a_lstm-python
```

39.5.17 Python LSTM notebook with plots

```
|-- Part4_NeuralFX
|    |-- 036a_lstm-python
```

39.5.18 C++ TorchScript LSTM

036b_lstm-torchscript

```
|-- Part4_NeuralFX
|    |-- 036b_lstm-torchscript
```

39.5.19 C++ LSTM processing a WAV

```
|-- Part4_NeuralFX
|    |-- 036c_lstm-wav
```

39.5.20 C++ LSTM performance test

```
|-- Part4_NeuralFX
|    |-- 036d_lstm-performance
```

39.5.21 C++ LSTM block-based processing

```
|-- Part4_NeuralFX
|    |-- 036e_lstm-blocks
```

39.5.22 C++ LSTM block-based processing with state

```
|-- Part4_NeuralFX
|    |-- 036f_lstm-blocks-memory
```

39.5.23 Python LSTM training program

```
|-- Part4_NeuralFX
|    |-- 037a_train_lstm
```

39.5.24 C++ trained LSTM JUCE project

```
|-- Part4_NeuralFX
|    |-- 037b_trained-lstm-JUCE
```

39.5.25 C++ trained LSTM RTNeural project

```
|-- Part4_NeuralFX
|    |-- 037d_lstm-rtneural
```

39.5.26 C++ trained LSTM RTNeural project

```
|-- Part4_NeuralFX
|    |-- 037e_lstm-rtneural-JUCE
```

39.5.27 Bonus project: C++ implementation of Steinmetz's steerable reverbs

```
|-- Part4_NeuralFX
|    |-- 038
a_neural_conv_steinmetz
```

Bibliography

[1] Andrew Maz. *Music Technology Essentials: A Home Studio Guide*. Focal Press, 2023.

[2] Ron Begleiter, Ran El-Yaniv, and Golan Yona. On prediction using variable order Markov models. *Journal of Artificial Intelligence Research*, 22:385–421, 2004.

[3] John A. Biles. Life with GenJam: Interacting with a musical IGA. In *IEEE SMC'99 Conference Proceedings. 1999 IEEE International Conference on Systems, Man, and Cybernetics (Cat. No. 99CH37028)*, volume 3, pages 652–656. IEEE, 1999.

[4] Jean-Pierre Briot, Gaëtan Hadjeres, and François-David Pachet. Deep Learning Techniques for Music Generation – A Survey, August 2019. arXiv:1709.01620 [cs].

[5] Noam Brown and Tuomas Sandholm. Superhuman AI for heads-up no-limit poker: Libratus beats top professionals. *Science*, 359(6374):418–424, 2018. Publisher: American Association for the Advancement of Science.

[6] Jatin Chowdhury. RTNeural: Fast Neural Inferencing for Real-Time Systems. *arXiv preprint arXiv:2106.03037*, 2021.

[7] Nick Collins, Vit Ruzicka, and Mick Grierson. Remixing AIs: mind swaps, hybrainity, and splicing musical models. In *Proceedings of the 1st Joint Conference on AI Music Creativity*, Sweden, 2020.

[8] Darrell Conklin. Music generation from statistical models. In *Proceedings of the AISB 2003 Symposium on Artificial Intelligence and Creativity in the Arts and Sciences*, pages 30–35. Citeseer, 2003.

[9] John Covert and David L. Livingston. A vacuum-tube guitar amplifier model using a recurrent neural network. In *2013 Proceedings of IEEE Southeastcon*, pages 1–5. IEEE, 2013.

[10] Giovanni De Sanctis and Augusto Sarti. Virtual analog modeling in the wave-digital domain. *IEEE transactions on audio, speech, and language processing*, 18(4):715–727, 2009. Publisher: IEEE.

[11] Nina Düvel, Reinhard Kopiez, Anna Wolf, and Peter Weihe. Confusingly Similar: Discerning between Hardware Guitar Amplifier Sounds and Simulations with the Kemper Profiling Amp. *Music & Science*, 3:205920432090195, January 2020.

[12] Jesse Engel, Chenjie Gu, Adam Roberts, and others. DDSP: Differentiable Digital Signal Processing. In *International Conference on Learning Representations*, 2019.

[13] Fiammetta Ghedini, François Pachet, and Pierre Roy. Creating Music and Texts with Flow Machines. In Giovanni Emanuele Corazza and Sergio Agnoli, editors, *Multidisciplinary Contributions to the Science of Creative Thinking*, pages 325–343. Springer Singapore, Singapore, 2016. Series Title: Creativity in the Twenty First Century.

[14] Gaëtan Hadjeres, François Pachet, and Frank Nielsen. DeepBach: a Steerable Model for Bach chorales generation. *arXiv preprint arXiv:1612.01010*, 2016.

[15] Dorien Herremans, Ching-Hua Chuan, and Elaine Chew. A Functional Taxonomy of Music Generation Systems. *ACM Computing Surveys*, 50(5):1–30, September 2018.

[16] Lejaren A. Hiller Jr and Leonard M. Isaacson. Musical composition with a high speed digital computer. In *Audio engineering society convention 9*. Audio Engineering Society, 1957.

[17] Sepp Hochreiter and Jürgen Schmidhuber. Long short-term memory. *Neural computation*, 9(8):1735–1780, 1997. Publisher: MIT press.

[18] Geoffrey Holmes, Andrew Donkin, and Ian H Witten. Weka: A machine learning workbench. In *Proceedings of ANZIIS'94-Australian New Zealnd Intelligent Information Systems Conference*, pages 357–361. IEEE, 1994.

[19] S. R. Holtzman. Using generative grammars for music composition. *Computer music journal*, 5(1):51–64, 1981. Publisher: JSTOR.

[20] Feng-hsiung Hsu. IBM's deep blue chess grandmaster chips. *IEEE micro*, 19(2):70–81, 1999. Publisher: IEEE.

[21] Shulei Ji, Jing Luo, and Xinyu Yang. A comprehensive survey on deep music generation: Multi-level representations, algorithms, evaluations, and future directions. *arXiv preprint arXiv:2011.06801*, 2020.

[22] Boris Kuznetsov, Julian Parker, and Fabian Esqueda. DIFFERENTIABLE IIR FILTERS FOR MACHINE LEARNING APPLICATIONS. In *Proc. Int. Conf. Digital Audio Effects (eDAFx-20)*, 2020.

[23] Yann LeCun, Yoshua Bengio, Geoffrey Hinton, Lecun Y., Bengio Y., and Hinton G. Deep learning. *Nature*, 521(7553):436–444, 2015. ISBN: 3135786504 _eprint: arXiv:1312.6184v5.

[24] Sungho Lee, Hyeong-Seok Choi, and Kyogu Lee. Differentiable artificial reverberation. *IEEE/ACM Transactions on Audio, Speech, and Language Processing*, 30:2541–2556, 2022. Publisher: IEEE.

[25] Joel Lehman, Jeff Clune, Dusan Misevic, Christoph Adami, Lee Altenberg, Julie Beaulieu, Peter J. Bentley, Samuel Bernard, Guillaume Beslon, and David M. Bryson. The surprising creativity of digital evolution: A collection of anecdotes from the evolutionary computation and artificial life research communities. *Artificial life*, 26(2):274–306, 2020. Publisher: MIT Press One Rogers Street, Cambridge, MA 02142-1209, USA journals-info . . . .

[26] Louis McCallum and Mick S Grierson. Supporting Interactive Machine Learning Approaches to Building Musical Instruments in the Browser. In *Proceedings of the International Conference on New Interfaces for Musical Expression*, pages 271–272. Birmingham City University Birmingham, UK, 2020.

[27] Eduardo Reck Miranda. Cellular Automata Music: An Interdisciplinary Project. *Interface*, 22(1):3–21, January 1993.

[28] David Moffat. AI Music Mixing Systems. In *Handbook of Artificial Intelligence for Music*, pages 345–375. Springer, 2021.

[29] Hans Moravec. When will computer hardware match the human brain. *Journal of evolution and technology*, 1(1):10, 1998.

[30] Gerhard Nierhaus. *Algorithmic composition: paradigms of automated music generation*. Springer Science & Business Media, 2009.

[31] Aaron van den Oord, Sander Dieleman, Heiga Zen, Karen Simonyan, Oriol Vinyals, Alex Graves, Nal Kalchbrenner, Andrew Senior, and Koray Kavukcuoglu. Wavenet: A generative model for raw audio. *arXiv preprint arXiv:1609.03499*, 2016.

[32] Andrew Pickering. Cybernetics and the mangle: Ashby, Beer and Pask. *Social studies of science*, 32(3):413–437, 2002. Publisher: Sage Publications London.

[33] Will Pirkle. *Designing audio effect plugins in C++: for AAX, AU, and VST3 with DSP theory*. Routledge, 2019.

[34] Nicola Plant, Clarice Hilton, Marco Gillies, Rebecca Fiebrink, Phoenix Perry, Carlos González Díaz, Ruth Gibson, Bruno Martelli, and Michael Zbyszynski. Interactive Machine Learning for Embodied Interaction Design: A tool and

methodology. In *Proceedings of the Fifteenth International Conference on Tangible, Embedded, and Embodied Interaction*, pages 1–5, 2021.

[35] MA Martínez Ramírez, Emmanouil Benetos, and Joshua D. Reiss. Deep learning for black-box modeling of audio effects. *Applied Sciences*, 10(2):638, 2020. Publisher: MDPI AG.

[36] Anders Reuter. Who let the DAWs Out? The Digital in a New Generation of the Digital Audio Workstation. *Popular Music and Society*, 45(2):113–128, March 2022.

[37] Curtis Roads and Paul Wieneke. Grammars as representations for music. *Computer Music Journal*, pages 48–55, 1979. Publisher: JSTOR.

[38] David Silver, Julian Schrittwieser, Karen Simonyan, Ioannis Antonoglou, Aja Huang, Arthur Guez, Thomas Hubert, Lucas Baker, Matthew Lai, Adrian Bolton, and others. Mastering the game of go without human knowledge. *nature*, 550(7676):354–359, 2017. Publisher: Nature Publishing Group.

[39] Christian J. Steinmetz and Joshua D. Reiss. Steerable discovery of neural audio effects. *arXiv preprint arXiv:2112.02926*, 2021.

[40] Christian J. Steinmetz and Joshua D. Reiss. Efficient neural networks for real-time modeling of analog dynamic range compression, April 2022. arXiv:2102.06200 [cs, eess].

[41] Fabian-Robert Stöter, Stefan Uhlich, Antoine Liutkus, and Yuki Mitsufuji. Open-unmix-a reference implementation for music source separation. *Journal of Open Source Software*, 4(41):1667, 2019.

[42] Fractal Audio Systems. Multipoint Iterative Matching & Impedance. Technical report, Fractal Audio, 2013.

[43] Tara Vanhatalo, Pierrick Legrand, Myriam Desainte-Catherine, Pierre Hanna, Antoine Brusco, Guillaume Pille, and Yann Bayle. A Review of Neural Network-Based Emulation of Guitar Amplifiers. *Applied Sciences*, 12(12):5894, January 2022. Number: 12 Publisher: Multidisciplinary Digital Publishing Institute.

[44] Thomas Wilmering, David Moffat, Alessia Milo, and Mark B. Sandler. A History of Audio Effects. *Applied Sciences*, 10(3):791, January 2020. Number: 3 Publisher: Multidisciplinary Digital Publishing Institute.

[45] Alec Wright, Eero-Pekka Damskägg, Lauri Juvela, and Vesa Välimäki. Real-Time Guitar Amplifier Emulation with Deep Learning. *Applied Sciences*, 10(3):766, January 2020. Number: 3 Publisher: Multidisciplinary Digital Publishing Institute.

[46] Alec Wright, Eero-Pekka Damskägg, and Vesa Välimäki. Real-time black-box modelling with recurrent neural networks. In *22nd international conference on digital audio effects (DAFx-19)*, pages 1–8, 2019.

[47] Matthew Yee-King and Martin Roth. Synthbot: An unsupervised software synthesizer programmer. In *International Computer Music Conference, Belfast 2008*, Belfast, 2008.

[48] Matthew John Yee-King. *Automatic Sound Synthesizer Programming: Techniques and Applications*. PhD thesis, University of Sussex, 2011.

[49] Matthew John Yee-King, Leon Fedden, and Mark d'Inverno. Automatic programming of VST sound synthesizers using deep networks and other techniques. *IEEE Transactions on Emerging Topics in Computational Intelligence*, 2(2):150–159, 2018. Publisher: IEEE.

[50] Nimalan Yoganathan and Owen Chapman. Sounding riddims: King Tubby's dub in the context of soundscape composition. *Organised Sound*, 23(1):91–100, 2018. Publisher: Cambridge University Press.

Index

Abstract class, 148
API, 15
Artificial Intelligence
 Definition, 10
Audio effects
 History, 210
 Types, 210
AudioPluginGraph
 Add a plugin, 139
AudioProcessGraph
 Initialisation, 137
AudioProcessorGraph, 135

Back propagation, 258
Batch, 96
Batch processing, 258
Batch size, 298
Black box models, 254
Blocks, 274
Book
 Aims, 3
 How to use it, 5
 Overview, 5
 Who is it for?, 6
Build tool, 11

CMake, 32
 Commands, 34, 35
 Description, 13
 JUCE, 40
 Libtorch and JUCE, 50
 Libtorch configuration, 47
 macOS install, 33
 Windows install, 32

Colab, 58
Convolution
 Frequency domain, 226
 JUCE, 227
 Time domain, 222
CUDA, 302

Dataloaders, 302
Dataset, 121
 Noise, 89
 Preparation, 292
 Storage, 121
DAW, 16
DBG for Windows, 27
Development environment, 11
 Components, 11
 Problems, 60
 Setup, 18
DLL, 15, 48
DSP, 214
 Impulse response, 216
 LTI systems, 217
Dynamic linking, 15, 48

Epoch, 96
Error function, 94

FIR, 214
FM synthesis algorithm, 72
Frequency response, 236

Generalisation, 301
Generating sequences, 171
Generative model, 15
GPU, 17

Guitar amp emulator, 249

Headers, 41

IDE, 13
IIR
    C++, 235
    Definition, 231
    JUCE, 237
    One pole, 233
IIR/FIR different equation, 235
Impulse response, 216
Inference, 14, 100, 259
    Performance, 320
Install libtorch, 44
Inter-onset interval, 188
Interactive machine learning, 100
Interpreted program, 14
IOI, 188
IOI Model
    Generate, 191

JUCE
    Add a slider, 66
    CMake, 40
    Installing, 24
    Linux, 29
    macOS, 27
    MIDI input, 69
    RTNeural, 323
    Standalone target, 26
    Why?, 23
    Windows, 26
JUCE CMake commands, 40

Libtorch
    Description, 44
    Install, 46
    Simple example, 44
Linear regression
    Neural networks, 85
    Synthesizer control, 82
Linux

Build, 39
Load a WAV file, 220
Loss functions, 305
LSTM, 262
    Audio, 265
    Description, 266
    JUCE, 282
    Model, 303
    Performance, 271
    Process a WAV, 269
    State, 276, 280
LSTM state and block size, 297

Markov model
    C++, 169
    Definition, 163
    Generation, 171
    Higher order, 165
    Observation, 163
    Polyphony, 204
    State, 163
    State transition, 163
    State transition matrix, 164
    Variable order, 166
Meta-controller purpose, 81
MIDI Chord detection
    C++, 203
MIDI input, 69
Moravec, 2
Musical agents, 160

Native program, 13
Neural effects
    Introduction, 212
Neural FX
    Reverb, 327
Neural network
    as FIR and IIR, 256
    C++ class, 109
    Definition, 256
    Finding best fit, 90
    Forward, 114

Linear model, 91
Linear regression, 85
Passing data to it, 108
Plugin control, 153
Training data, 154
Note duration, 198

Optimiser, 95
Over-fitting, 119, 301

Parameters, 77
Pitch model, 176
Generating, 180
Plugin example
Sinewave synth, 62
Plugin host, 29
Plugin hosting
Load a plugin, 131
Plugin locations, 30
Plugins
Adding parameters, 77
FM synthesizer, 72
Hosting, 129
Lifecycle, 64
Parameters, 151
processBlock, 65
Showing user interface, 145
processBlock function, 65
Projucer, 24
Python
Installing packages, 56
Linux install, 54
macOS install, 54
Neural FX setup, 262
Notebooks, 56
Training setup, 289
Virtual environments, 55
Windows install, 53

Quantisation, 199

Rhythm, 187
RTNeural, 320

Exporting weights, 321
JUCE, 323
Performance, 323

Sequence modelling techniques, 162
Sequencers, 158
Signals, 214
Sliders, 66
Smart pointers, 123
Softmax, 115
Source code repository, 8
Superknob, 104
Systems, 214

Tensor, 93, 264
Tensors
C++, 93
Torchknob, 107
TorchScript
Exporting, 267, 315
Importing, 268
Trained model, 14
Training, 96, 259, 287
Definition, 256
Example runs, 313
Loss, 305
LSTM, 312
UI, 119
Validation and test data, 301
Training loop, 122, 127
Training, definition of, 255

Validation data, 301
Variable order Markov model, 166
Velocity model, 205
Virtualenv, 55
Visual Studio
Installing, 19
Visual Studio Code, 21
CMake build, 36
Installing, 20

Waveshaper

C++, 242
Definition, 241
JUCE, 243
Parameters, 247
Wekinator, 99
Inputs and outputs, 100
Windows
Visual Studio, 37
Windows debug output, 27

Xcode
Build, 39
Installing, 19